COPPER FOR AMERICA

Other Books by Maxwell Whiteman

A Century of Fiction by American Negroes, 1955
A History of the Jews of Philadelphia from Colonial Times to the Age of Jackson (with Edwin Wolf 2nd), 1957
Mankind and Medicine: A History of Philadelphia's Albert Einstein Medical Center, 1966
Pieces of Paper: Gateway to the Past, 1967
While Lincoln Lay Dying, 1968

EDITOR:

Afro-American History Series, forty-six volumes, 1969
The Kidnapped and the Ransomed: The Narrative of Peter and Vina Still After Forty Years of Slavery, with an introductory essay on Jews in the antislavery movement, 1970

COPPER
FOR AMERICA

The Hendricks Family
and a National Industry
1755-1939

MAXWELL WHITEMAN

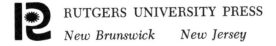

RUTGERS UNIVERSITY PRESS
New Brunswick   New Jersey

Copyright © 1971 by Rutgers University, the State University of New Jersey

Library of Congress Catalog Card Number: 79-153446

ISBN: 0-8135-0687-5

Manufactured in the United States of America by Quinn & Boden Company, Inc., Rahway, New Jersey

For
Bethel Delano Whiteman

Harmon Hendricks, 1771–1838. *Portrait by S. L. Waldo and William Jewett, courtesy of the Frick Art Reference Library.*

Preface

For many years the papers that comprise the Hendricks collection were looked upon as a treasure in Jewish history, but not as a source of early American technology or economic history. No one who has examined this collection can doubt its importance for the study of Jewish life in the United States, yet it is difficult to believe that the ledgers, daybooks, cash books, invoices and the records of ship coppering were viewed with disregard by a school of parochial historians. It is a tribute to the late Henry S. Hendricks who thought otherwise about the early carrying trade and the development of the nonferrous metal industry. At his suggestion, and with the object of incorporating both aspects of the family's activity, this book was written.

When the collection came to my hands it was in a state of disorganization. Its present arrangement is largely the work of Elizabeth Delano Whiteman. Her knowledge of the first three decades of the nineteenth century contributed to the valuable manuscript guide that was prepared. Those who use the collection in the future will be as grateful to Mrs. Whiteman as I was in studying and absorbing the details of the Hendricks firm and family life.

My gratitude to Helen R. Hendricks is greater than I can reveal. Her extensive knowledge of the family past provided me with a perspective that is not to be acquired through the study of manuscripts. It is a matter of deep regret that Miss Hendricks died before this book appeared. To her sister Ethel Hendricks

Frank I am indebted for the subtler forms of guidance, and to Rosalie Nathan Hendricks who spent endless hours in making available sources of information that were not otherwise obtainable, I am particularly appreciative. Congregation Shearith Israel, the oldest Jewish house of worship in the United States in which the religious history of the Hendricks family can be traced from colonial times to the present day, graciously permitted me the use of its valuable archives.

Dr. James J. Heslin, Director of the New York Historical Society where the Hendricks collection was deposited, stepped beyond professional obligation to be of help and I use this occasion to express my thanks to him. Mr. Gordon Wright Colket of Gladstone, New Jersey, brought to my attention and made available a large collection of Hendricks-Tobias papers. Without them much of the activity of the Philadelphia manufacturers would have remained obscure. I have leaned heavily on the work of Professor Bradford Perkins and his influence is implicit for the years before the War of 1812. Professor Harry H. Shapiro of Rutgers University provided valuable guidance for which I am indebted.

Both Ann Rosenberg and Evelyn Weiman of Philadelphia were helpful in making suggestions which were incorporated in the manuscript.

It is a matter of personal regret that the final chapter was reduced to a brief epilogue. If the documentary materials for the last quarter of the nineteenth century and the early decades of the twentieth century have survived and can yet be located, an important chapter on the history of the Hendricks mill at Belleville awaits to be written.

Maxwell Whiteman

Elkins Park, Pennsylvania
February 1971

Contents

COPPER FOR AMERICA

1

URIAH HENDRICKS

Colonial Tradesman and Ironmonger

I

The history of the copper trade prior to the Civil War is so fragmentary that the story of the Hendricks family contributes extensively to our knowledge of the American copper industry. There were few mines and no rolling mills in the American colonies when the metal business that carried the Hendricks name from 1764 to 1939 was founded by Uriah Hendricks. The copper that was mined in colonial America was exported to England to be used by her mills and industries. The colonial merchant had no opportunity to enter into copper manufacturing; he could only import manufactured copper. Copper imports quickly became part of the ironmonger's business.

After the American Revolution copper sheathing was used for the protection of wooden ships. More than a decade passed before this innovation attracted the attention of the United States Navy. The practical and clever shipping merchants preceded the Navy in adopting copper sheathing to protect the ships that brought their goods across the Atlantic.

Without an adequate native source the United States was dependent upon England and South America for copper and other nonferrous materials. Copper procurement was severely affected by European wars at the beginning of the nineteenth century. A cursory examination of the Hendricks records indicates that this phase of the carrying trade, before the War of 1812, has not received adequate treatment by economic historians. Harmon Hendricks, the second of the family to continue in the trade, became the dominant figure in the copper and tin market prior to the war with England and he rose to a position of prominence in the first quarter of the nineteenth century. In 1813 he reclaimed the deserted Soho Copper Works in Belleville, New Jersey, on the site of one of the oldest American mills, close to the Schuyler copper mine. Here, where his predecessors, Nicholas J. Roosevelt and Robert R. Livingston, had failed, Hendricks was successful in manufacturing and rolling copper.

Harmon Hendricks' transactions with Paul Revere and Robert Fulton, with the engineering companies of James B. Allaire and the West Point Foundry Association, and with all the shipbuilders of New York City, further enrich our knowledge of copper in this "age of steam." The record of the packet boats, the steamboats, the whalers, the clipper ships that reached the China seas, and the vessels of the United States Navy, which were coppered by three generations of the Hendricks family, is drawn from the collection of family and business papers preserved by the family.

For more than five generations the Hendricks name was associated with the American copper trade; it is the oldest name in the industry before World War II. And for more than two and a half centuries the family and its colonial antecedents have been identified with the civic, cultural, and religious life of New York City.

Uriah Hendricks, who pioneered in the copper trade, came to New York in the fall of 1755. He was one of a number of young, single Jews, encouraged by their relatives, to undertake the long voyage to the American colonies during the 1750's. Those with established business connections, provided with a small working capital, and assured of a supply of English goods were relieved of some of the difficulties that confronted many colonial immigrants. Those who could turn to a friendly fellow countryman for aid during their first days in the colonies lost no time in obtaining a foothold.

Uriah Hendricks was trained in his father's London counting house, where he had shown such an aptitude for the colonial trade that his father urged him to go to New York and establish a business of his own. The prospect that such a colonial outlet could be supplied by the London house enhanced the proposition and, to ease the condition of Hendricks' settlement in New York, letters of introduction were given to Jacob Franks, the head of an outstanding firm of colonial merchants who maintained business houses in London and New York.

Stocky, dark-haired Uriah Hendricks, not yet nineteen years old, began business modestly on lively Hanover Street, whose shops extended from Wall Street to Pearl Street. His shop, a choice location next door to the ironmongery of Goelet and Curtenius, was easily recognized by the sign of the Golden Key, which hung over the ironmongery and cast its shadow over the Hendricks front. Opposite the Golden Key was John Dies' ironmongery and close by was the dry goods store of Naphtali Hart Myers, a business associate of the London Hendricks. Not far from Hanover Street were the establishments of the wealthy merchants and shipowners. Their numerous employees were the object of envy of the small shopkeepers, whose limited resources compelled them to live above their crowded shops and to conduct their businesses with family help. Only when better times permitted were they able to afford a servant or an apprentice. Hendricks lived away from his shop, and as a young beginner he could not afford a hired hand.

Hendricks sold popular articles at prices that attracted the fashionable New York gentry, such as inexpensive jewelry, trinkets, and select dry goods. The typical wares of the colonial shopkeeper that were constantly in demand he sold for ready cash. Gradually, "looking glasses," "Clocks that strikes every half Hour, Repeating Watches," and "goods fit for the Season" were added to his small stock. Though such merchandise did not compete seriously with the goods offered by the oldtimers, whose imports arrived by the same London ships, Uriah Hendricks soon was able to bring his name into association with the established merchants and tradesmen by advertising in the New York press. With each new arrival of goods his advertisements found a prominent place alongside those of his older and more experienced competitors. His relationship with the men of business whom Hendricks came to know was strengthened by their association in the Spanish-Portuguese congregation, Shearith Israel, located on nearby Mill Street. Here he had the opportunity to make friends with some of the younger, New York-born Jews. An interest in metals brought him in closer contact with the Myers brothers: Asher, a copper brazier; Myer, a goldsmith; and Joseph, a silversmith. The older Jewish families of the city, particularly the Gomez and Seixas families, who took an active part in the affairs of Shearith Israel, eyed the young man with favor. Uriah Hendricks was observant in religion and well versed in the Bible. Although accustomed to the ritual of the Great Synagogue in London, he readily adapted himself to the Spanish-Portuguese custom of New York. Before long, in the fall of 1756, he was accorded a choice synagogue honor: he was the first to bless the reading of the Pentateuch at the beginning of its annual cycle, and thereafter his name appears frequently in the rolls of the congregation.

Sometimes on a Sunday afternoon he would spend his leisure time horseback riding, or chatting with the tradesmen over a cup of tea. Yet despite these pleasant pastimes and the company of his new friends and business associates, his letters back home revealed loneliness. To his brother Abram he wrote that nothing would please him more than to "embrace your presence & have the pleasure of a brother, friend & a companion in this remote place." Abram did not accept the invitation to come to America, and Uriah continued to live alone in modest rented quarters. "As their is no houses to be had in any good place of this city therefore [it] is a great favour that I continue for £30 . . ." a year. The cost of living had risen since he had come to New York, "everything in the way of living being dearer than in London."

It soon became clear to those who knew Uriah Hendricks that he was energetic, quick to learn, fair in all of his dealings, and showed a grasp of business conditions that equaled that of older and more experienced men. As far as he was able Uriah remained aloof from the petty controversies that involved many of his fellow businessmen. It was difficult enough to get a foothold in business without becoming entangled in the quarrels that arose from competition, and Uriah confined himself to expressing his feelings about the clashes of the New Yorkers in his letters back home.

A little more than a year after his entry into the Hanover Street trade, business had improved sufficiently for nineteen-year-old Uriah to move to larger quarters, and in less than a year and a half after that, he opened his third shop. Every move was one of advancement. By a series of coincidences, each time he became the neighbor of an ironmonger. To his right was John Troup's ironmongery and to his left, two doors away, was the printing office of Hugh Gaine, publisher of the *New York Mercury*, in which Hendricks advertised his imports of "European and East-India goods."

Hendricks' connections with the Frankses, an important mercantile family, were of considerable value to his business. From London, Moses Franks supplied him regularly with an assortment of goods, negotiated his bills of exchange, and represented him in his dealings with various English exporters. Uriah's main source of supply, however, was his father. Aaron Hendricks kept his son informed of the fluctuations of the English market. Uriah, in return, advised his father of American needs. Woolens, for example, were in good demand and were shipped in the spring in time for sales in the fall. Old clothes were not desirable for—he wrote his father: "People here are to proud to wair second hand clothes." In addition to the woolens and other dry goods, the glass, silverware, and copper nails that were supplied by his father, and the lesser items purchased from his brother Abram, who had also ventured into business, Hendricks handled merchandise from his uncle Harmon and other London relatives. His oldest sister, Dalia Pollock, would occasionally consign to him a few articles of jewelry or silverware from her husband's business, and these small transactions, if not very profitable, gave Uriah great personal satisfaction.

A man of quite a different character was Uriah's brother-in-law, Levy Cohen, a hard-driving businessman who fancied himself a person of importance. Though Uriah learned to avoid the business harangues of the New Yorkers, he could not easily avoid a brother-in-law who was suspicious of every transaction across the Atlantic. Levy Cohen insisted upon receiving prompt payment when long-term negotiations were involved. Merchandise of inferior quality shipped from Cohen's London warehouse could not be returned, nor could it be sold for prices below those fixed by the consignor; but Hendricks, by means of hard work, met Cohen's demands. Gossip concerning the steady flow of goods from Moses Franks to Hendricks irked Cohen, who felt that he and the rest of his family were thus being deprived of business. Cohen complained constantly to his father-in-law who, as the mediator, persuaded Uriah to overlook Cohen's foibles. Echoes of Cohen's insults even reached the New York trade.

When the complaints were further compounded by abusive letters, Hendricks finally declared himself to Cohen: "In short my caracter is so well establisht that I defy you to blot it anyway." Cohen was unconvinced by his brother-in-law's attempt to explain the slower methods of trade in the colonies, and ignored Hendricks' assertion that his reputation as a tradesman in New York was beyond reproach. Perhaps he underestimated Hendricks' sense of dignity, for the insulting letters continued. Obviously perturbed, Hendricks showed himself as superior to the insensitive Cohen by responding that he "never could have Imagin a man of learning could be guilty of such insults." To emphasize his feelings, he added a precept in Hebrew, drawn from Ecclesiastes, which he combined with the opening words from a verse in Proverbs, "Better is a dry morsel with quietness than both hands full of travail and the vexation of spirit." The peevish Cohen again turned to Aaron Hendricks, Uriah's father, but Uriah was so strongly provoked that he determined to have no further business with Levy Cohen.

Other family contacts were pleasant. To his two married sisters he occasionally sent a thoughtful gift of a cask of pickles, some anchovies, or a piece of silverware. His unmarried sisters, Rosy and Hannah, who also lived in London, enjoyed his solicitude, which was frequently expressed in a sentence or two in letters otherwise full of quotations of prices current or accounts of the latest ship arrivals from England. His interest in Rosy went beyond ordinary sibling obligation when there was some thought of her coming to New York to seek a prospective groom. In the spring of 1758, Uriah wrote his brother Abram that he

Should be Glad to hear of our sister Rosy being well settled Even in New York. I thought was once near making an advantageous Match for her. I made an offer of £400 . . . but £1000 was Demanded but then it was to our Present *parnass* [president] of the Kehillah, Batchelore of 34 years of age whom I suppose to be worth £3000 stg. My endeavors for her advantage will nevertheless not be wanting.

As a matrimonial matchmaker Hendricks was not successful. Rosy never came to New York, but later married Moses Oppenheim of London. The unnamed president of the synagogue was Samson Simson, a prosperous New York merchant and shipowner, who remained a bachelor to the end of his days. Hannah, Uriah's other sister, married a London Mordecai in the summer of 1758. Since Uriah had closed his business account with Levy Cohen, his two new brothers-in-law sought his favor in the trade across the Atlantic.

While he was defending himself against the hostilities of Cohen, Uriah Hendricks became involved in what was to be the most exciting aspect of his experience as a colonial tradesman— supplying the sutlers on the New York frontier during the French and Indian War. The fighting between the French and the English for supremacy over the colonies fired Hendricks' imagination. News of the war's progress was reported in Hugh Gaine's *Mercury*, which Hendricks read as soon as he could purchase it from the publisher next door. Accounts of the Indian depredations in the press corresponded to those received from the Jewish sutlers, whom he was supplying at the army posts.

At the busy Hanover Street shop Uriah Hendricks made up the loads of goods, filled the casks, and packed the bundles that were then sent up the river to Albany, reloaded, and hauled by ox carts inland. Long days at the shop were required to keep the sutlers well stocked. Although he never ventured out of the Hanover Square district, Uriah was personally and financially involved in the war; he felt a deep personal concern for his friends, the safety of his goods, and the progress of the conflict.

Indian massacres imperiled the lives of the sutlers, whose pack trains were frequently pillaged. Inevitably the brisk trade at the outposts came to an end because of the constant danger from both Indian and white marauders. Backwoodsmen themselves

sought safety within the army forts or fled to the East. Only a
few were daring enough to remain exposed to the Indian attacks.
Of the Jewish sutlers who were caught in the midst of the conflict, Benjamin Lyon, known as "little Labe," one of Hendricks'
ship companions to America, was at Fort Henry, and another
sutler and business account of Hendricks, Manuel Josephson, was
at Fort Edward. Josephson witnessed the arrival of the survivors
of the Fort Henry massacre, and Benjamin Lyon's coming "in safe
amongst the rest but much fatigued." Lyon's harrowing experience during the Indian attack at Fort Henry, however, did not
deter him from continuing to trade with the Army. A year later,
in the summer of 1758, Uriah wrote to his father that "little Labe"
was captured by the French on his way to the British camp at
Lake George. About the same time, Lyon's partner, Gerson Levy,
Hendricks' other ship companion, was captured by the Indians.
Lyon managed to slip through the French cordon and Levy
escaped from the Indians. Both men stuck to their hazardous
trade throughout the French and Indian War and later led in
forming a band of war-toughened Jewish sutlers, who finally
migrated to Canada.

In closing his account with his brother-in-law Levy Cohen,
Uriah was compelled to call upon his sutler friends to collect
some debts for Cohen. In a letter to Cohen he explained why
Americans were "slow-pay":

Your favour of the 20th March came Duely to hand Inclosing a Bill of
Exchange in Your favour on Messrs. Solomon & ComY which Bill I got Little
Levy to Except here the 25th Ultimon. As his post men is among the Army
at Oswego three hundred Miles thru the woods Beyond Albany at wi^{ch} place
It was not prudent to send your Bill for Exceptance & Levy is gone to
Albany & promest me to Remitt a Bill therefor by the time Due if tew
months after It is Reckoned Good pay here. Your Acct.^{s} of Ballance Cant
bee settled yett as have mett with Disapointments in Urging some old
Standing Debts of yours &c perticulerly of Asher & Saml Isaacs who have
stopt payment & owes £ 200 currency of which I Believe £ 30 or thereabout
is on your acct: But the Greatest part of which is on my acct: I am flattered
they have enough to pay the whole But am afraid however in a few Days
I shall know . . .

Before he came to America Benjamin Lyon had apparently
known Uriah's Aunt Hava, for there was some talk of a marriage

between them. Uriah's familiarity with the life of the army sutlers
with whom he did business enabled him to advise his father about
the desirability of Lyon as a prospective suitor for Hava, in the
same way that he had advised his brother about Samson Simson
earlier:

As for Little Labe & Co.ʸ they be worth money as they Trade Considerably
with the Army But make all that when they Continually Break *Shabbat* &
Eat *trefot* [ritually forbidden food] & have no Regard to Religion. I have
such a Regard for my Ant *Hava* would as Soone Loose £ 50 than to hear
of his having her. Altho I trade with them as they make tollerable payment.

Despite the many risks involved Hendricks found trading with
the sutlers profitable. They were trustworthy, they paid promptly,
and they successfully eluded danger. However, many sutlers disappeared in the wilderness or absconded with the goods consigned to them. They constantly faced the unknown hazards of
the forest. If they survived physical hardship, they were often
eliminated from business because their credit, meanwhile, had
been damaged. Hendricks had complete faith in the two sutlers
to whom he sold goods worth hundreds of English pounds. They
corresponded regularly and the intimate tone of their multi-
lingual correspondence reflected their friendship. Hendricks often
would transliterate an English phrase into Hebrew or include a
passage in Yiddish describing the scarcity of goods. Thus Levy
and Lyon could better understand his point, and the prying draymen, unfamiliar with these languages, were kept from obtaining
information concerning the price or value of the goods. Hendricks looked with disfavor on the sutlers' disregard for religion,
but this did not diminish their friendship or influence their business relationship. Hendricks did not hesitate to call on Levy to
collect a bill from the inland tradesmen of New York, and he
returned the favor by giving Levy a letter of credit, to be used
on a London trip, which emphasized Levy's good business reputation in the colonies.

In contrast to Lyon and Levy, who seemed to thrive in the
wilderness, scholarly Manuel Josephson, having encountered the
survivors of one Indian massacre, thereafter preferred the safety
of the city. As an observant Jew he desired to be nearer to the
synagogue. Back in New York, Josephson took up shopkeeping

"near the Slip-Market" where Hendricks first entered business, and the two men renewed a friendship that lasted their entire lives.

II

Although the business of supplying army sutlers occupied much of Hendricks' time, the war at sea also drew his attention. The colonial shipowners who carried his goods across the Atlantic were turning their trading vessels into privateers—armed cruisers that went to sea to prey on enemy shipping. Such prizes as they captured they redeemed for cash and choice merchandise. Shipowners shared these profits with captain, crew, and investors in the privateer. Privateering as a profession came into vogue prior to the French and Indian War but grew in volume during the war because shipowners were attracted to a lucrative field of business. The British sanctioned it because it hurt competitive shipping, and it enabled their navy to be deployed elsewhere.

The hazards of privateering were great, but a successful privateer had the opportunity to reap a fine profit. Precisely how Hendricks, at the age of twenty-one, was induced to invest in privateers is not known. Perhaps he was influenced by Hayman Levy, who outfitted the *Dreadnaught* and sent her to sea, or by Samson Simson, whose four vessels were also sent out as privateers. Both Levy, the largest direct trader with the Army, and Simson, a successful merchant in the carrying trade, were men of experience whom Hendricks might have attempted to emulate. Or perhaps he was influenced by the ship captains with whom he did business.

Whatever the source of his interest, Hendricks turned to Captain Robert Troup, a professional privateersman and master of the *Sturdy Beggar*. By the end of 1757 Troup had been warmly praised for his fearless and profitable expeditions. His *Sturdy Beggar*, "a ship carrying 26 guns and credited with a complement of 200 men," was a ship in which Uriah Hendricks willingly invested.

Captain Troup turned his attention to the French privateers, who had successfully harassed shipping on the British coast and were now looking for new challenges in the waters of colonial America. Troup's targets also included Spanish and Dutch ships bound for the West Indies in normal trade or privateering expeditions which hoped to take advantage of the war between England and France. This wartime privateering was a way of striking back at enemy privateers and of retrieving some of the colonial losses.

In the spring of 1758, after the return of Troup's swift-sailing cruiser, Hendricks described his first venture into privateering to his sister, Dalia Pollock:

Knowghing you rejoyce in my wellfair cannot omitt mentioning the little success have had in too private ships of war am concerned in, my have taking Prize. I suppose the amount of about £3000 currency in all Ster. Prize ship to, they took is condemd.

The success of this, the first of his recorded ventures, was not duplicated in later expeditions. The *Sturdy Beggar,* which had been successful in the past, only captured small prizes thereafter, but news of its exploits were promising:

Captain Troup sent in here in company with 3 other Privateers a small ship loaded with provisions may be worth about £2000 & by his letter of the 17th June He writes that he was then in Com.ʸ with 3 privateer ships of this port of 18 guns Each & a Briganteen of 24 guns off the west Corcoses awaiting for a Fleet they had infermation of & expected hourly of Nine Shugar ships under convoy of Tou Frigates of 30 guns Each bound from Cape Francoise for France which they are Determined to attack . . .

When Captain Troup finally arrived in New York, he reported only that, "fortune frown'd on him for he did not meet the above mentioned fleet." Hendricks feared that the Dutch prizes Troup had captured would be sent on to Jamaica to be recovered by a court of appeals, thereby reducing his potential profits. In the case of another prize taken by Troup, the appraised value was £1500 instead of the estimated £2000, and Hendricks further suspected that the whole affair would be enveloped in controversy. Even the lesser sum had to be shared with two sloop privateers. Before the Jamaica prize was settled Hendricks was busy outfitting the *Sturdy Beggar* anew, supplying most of the rigging himself. At the same time he optimistically purchased an interest in another privateer, the *Duke of Cumberland*.

To prepare himself for exigencies required in outfitting privateers Hendricks ordered up to a ton and a half of rope from his father for immediate shipment. He also offered his father a share in the *Duke of Cumberland*, in addition to an interest in the *Sturdy Beggar*. It was his intention, in the event that no prize was brought in by the *Duke of Cumberland*, to dispose of his interest in both ships.

Rumors of impending war with Spain further enhanced the prospects of the privateers, whose backers had grown anxious about the dull state of business prevailing in the colonies in 1759. Uriah, who had had fears about the *Duke of Cumberland*, found that they were justified. Her prize was poor and her management inefficient. He was also critical of the manner in which the shareholders went about dividing the spoils to their advantage, "and ganev [steal] what they please." He withdrew his interest in the ship at a loss. When the *Duke of Cumberland* was recommissioned on October 22, 1760, at 160 tons, 16 carriage guns and a crew of 50 men, it was in the name of another Jewish merchant, Judah Hays. As for the *Sturdy Beggar*, for which Uriah advanced £100 in rope and cash, she, too, was well equipped, with nine-pounders on her main deck and fours on her quarter deck. Hendricks placed much faith in her and was hopeful that her cruise would be more profitable than that of the *Duke of Cumberland*.

Financially Hendricks had considerable cause for concern. He daily expected to receive a £1000 shipment of fall goods from Moses Franks, with whom he had just placed a much larger order for summer goods for the following season. A bill from Franks covering goods purchased for the previous winter had recently fallen due. In addition he also awaited a substantial shipment from his father. And, finally, Uriah was anxious to settle his accounts with Levy Cohen.

Even after General Wolfe's Canadian victory over Montcalm on the Plains of Abraham business continued to slump. The merchants and tradesmen of New York spiritedly competed with each other for the sale of preferred items. Imports such as Russia goods and ravens duck, Indian blankets, ratteens, broadcloths, cambrics, oznabriggs, and harnesses could be sold for ready cash or three months' credit and, in a practical sense, these were choice peacetime goods. The market for other European items was slack, and only the tempting wheel of fortune, buying and selling lottery tickets, invited continuous speculation. When Captain Hunter's ship the *Leopard* arrived from London with a cargo of the more desirable merchandise shipped by Moses Franks, the two chief consignees were Uriah Hendricks and Judah Hays. With Hendricks' available cash invested in privateers, a quick turnover of these goods was even more important than before. In July of 1759 the new wares were advertised in the press.

Hendricks also urged his father to send more "Warlike stores," which had come into great demand. It was this type of goods —muskets, cutlasses, lead shot, powder, swivel guns and even six-pounders—that Samuel Judah and Hayman Levy were frequently advertising. Cordage, on the other hand, had been shipped into New York in such great quantities that it was "now Nearly a Drug." Furthermore, Samuel Judah was underselling Uriah Hendricks in cordage, but Hendricks firmly "resolved to sell at no such price" as Judah. He showed little fondness for Judah, whom he described to his father as one who

carried on considerable commerce here. Cant say the least in favour of his relations here neither of his strictness in our religion which are points to be considered . . . My father will please not to notice anything concerning Samuel Judah to anybody. As for my part I cant at all recommend him he being such a double tung'd chap.

On November 12, 1758, Captain Troup finally set sail in quest of prizes to compensate his efforts and reward the outfitters of the *Sturdy Beggar*. At the same time the French were seeking every opportunity to strike back at the American privateers. Occasionally a lone French ship would visit the coast, snatch its prey, and speed away. In one instance several swift-sailing French privateers swept the nearby coastline and withdrew with twenty American prizes in their possession. One of these prizes may have been the vessel belonging to Captain Howe, who was required by an admiralty court to sail for Antigua in order to ransom it, and concerning whom Uriah reported to his father that nothing further was heard after his departure from New York.

When the *Sturdy Beggar* returned in December some five weeks later, the report of her expedition was disappointing. Her prize was submitted to a court of admiralty for disposition. Now, more than ever, Hendricks was determined to sell his interest in the ship. New opportunities were beckoning. Sugar and coffee, highly desirable commodities, were being offered by the non-English islands of the West Indies, and all that he required was money and a responsible partner to share the risk. Cash was convincing in the French and Spanish ports of the West Indies and the trade was profitable. Monte Cristi had tons of sugar that it was eager to barter or sell. Only vessels sailing as neutrals were allowed to enter the trade, and Hendricks offered one of his London customers, Aaron Norden, the opportunity to join him in buying or building a vessel for the sugar trade. Young Hendricks was "determined to be an adventurer," eager to involve his father, brother, or Norden in the profitable commerce of the West Indies.

While these negotiations were in progress Captain Troup returned from still another ocean foray "Loaded with Shugar & Cofffee Likewise a Large Dutch Ship Loaded with the Same." Captain Troup had redeemed himself, and Uriah managed to make a small profit from his share of six casks of coffee in addition to the receipt of £50. But the total prize to be divided, part of which he could share with his father, was £1600 in coffee and sugar. Troup's success was propitious, for Hendricks had lost about £100 in the sale of his interest in the *Duke of Cumberland*, and domestic trade had continued to be indifferent. Although Uriah now had more cautious expectations of the privateers he had been depending on the *Sturdy Beggar* to recoup his losses. Troup took to the seas again, but this time lost sight of the fleet he was hunting. Instead he was drawn into His Majesty's service off the coast of Guadeloupe where he plundered "23 cask of sugar & 2 Negro Slaves Besides 100 French prisoners, with which he returned to Antigua."

The response from Uriah's relatives to the West Indies sugar venture was indifferent. He looked about for another partner and found one fairly quickly, Abraham Sarzedas, a tradesman eager to establish himself in the carrying trade. In the summer of 1759 Sarzedas and Hendricks prepared a small cargo of "sundry merchandize" to be shipped on board the *Five Brothers* to Monte Cristi, to be sold by the captain. The proceeds were to be invested "in good white sugar."

The transaction must have been successful, for a month later Uriah wrote his father of another venture to Monte Cristi. "Three days ago the Schooner Nancy Captⁿ Donellson saild from hence in which I own one fourth of the vessel & Cargo cost £800 currency." Hopes for the future success of the *Sturdy Beggar* gradually diminished. "The *Sturdy Bager* is now Layd Up for a Little time to see if times Alter In privateering which is now very Dull . . ." Despite his bad run of luck Captain Troup made another endeavor to outfit his ship with another merchant, Uriah Hendricks now being safely on the sidelines.

Success had not alienated young Hendricks from his family. His desire to return to England to see them was greater than the lure of uncertain profits in privateering. After he learned that his brother Abram would not come to New York his longing for his family increased. An attempt to encourage another relative to join him in the colony also failed. In letters to his family he declared his intention to return to London no less than ten times in as many months. To Abram he confided "that a lady of £500 currency would readily accept my address of a goodly family & good parts but no money will tempt me against inclination." At the age of twenty-two, after a four-year residence in New York, Uriah decided to return to London. Perhaps his disappointment in not finding a bride, coupled with the strong desire to see his family, persuaded him to make a voyage to England. In the fall of 1759 he promised himself that he would buy only one cargo for the coming spring, and then he planned to settle his business affairs in the colony.

It was no easy task for Uriah to extricate himself from a business in which he was deeply involved. He was still active in the Monte Cristi sugar trade and, as he wrote of his inventory, "at Present I suppose I have in store £2600 currency in goods." This stock required liquidation, commitments to the sugar trade had to be fulfilled, old bills had to be collected, and his small contribution to the building of the Newport synagogue had to be paid. Uriah estimated that it would require "18 or 20 months to settle" his colonial affairs.

Hendricks' spring cargo consisted of merchandise unlike that which he had previously imported: "A Parcel of choice London-made Men's and Boy's Shoes and Pumps. Likewise, Muscavado Sugar by the Hogshead." Throughout 1760 he supplemented his dwindling stock with small shipments of English goods that could be sold for ready cash or short-term credit, and he disposed of his remaining lottery tickets. Finally, in October, 1760, Hendricks announced in the New York press that he "intends next Spring for London, and earnestly requests all those in arears to him, to make speedy Payment."

In six years Uriah Hendricks ran an astonishing gamut of colonial trade. He had begun as an ordinary shopkeeper at the age of eighteen, and he emerged as a merchant by the time he was twenty-three. He had taken advantage of the import trade that rose steadily during the French and Indian War; he had speculated in privateering and sold his interest before he suffered serious losses; and his decision to return to London came at the beginning of the colonial depression of 1759.

Within the period of time of his first American residence the population of New York had steadily increased. A change in the ethnic background of the Jewish immigrants became evident in the mid-1750's, when Jews from Germany gradually came to outnumber those of Iberian descent. Many of these immigrants had passed through the counting houses of London, preparing themselves for careers in the American colonies. Quietly, Hendricks had surpassed his contemporaries. Energetic Barnard Gratz, who had come to Philadelphia in 1754 to enter the employ of David Franks, was still dreaming of becoming a merchant in 1759, when Uriah had already assured himself a reserve of £1000 for the voyage home. In the spring of 1761 Uriah set sail for London, uncertain whether he would ever return to New York.

III

Less than a year after Hendricks returned to London he brushed aside any doubts he might have had of a future in America and returned to New York. No record of his brief London residence has been located that might explain his decision to continue as a merchant in the colonies. If the lure of colonial trade was as strong as he indicated at the time when he was torn between seeing his family and continuing his business, it was reason enough to motivate his return.

Back in the now familiar surroundings of the old Battery, he located his new business at the lower end of Smith Street near the Sloat, behind Hanover Square and not far from the Gomez household. Except for the introduction of street illumination the city had not changed at all. The end of the French and Indian War appeared to be in sight, and the decline in business that had begun in 1759 had not yet halted. Many of Hendricks' former business associates were faced with the transition from wartime trade to the sale of peacetime goods. Uriah was spared the experience of this transition because of his absence from the country in 1761 and the liquidation of his business. There were murmurs among a few of the Jewish tradesmen of returning to London, and some who were unable to meet the new exigencies fell into bankruptcy. Hendricks at the age of twenty-five began anew, but as an old, experienced hand.

His reappearance at Shearith Israel was a welcome one. Old friends, Manuel Josephson, Jacob Franks, and the members of the Gomez family, gave him a kind reception. It was not unusual to find Uriah Hendricks as a guest in the socially prominent Gomez household. The Gomezes, merchant venturers who had come to New York from the British West Indies at the beginning of the eighteenth century, were also leaders in the Spanish-Portuguese congregation. It was Eve Esther, the daughter of the late Mordecai Gomez, to whom Hendricks turned for a bride; in the spring of 1762 nineteen-year-old Eve Esther was betrothed to Uriah Hendricks and they were married on June 30. Uriah settled the handsome sum of £1000 on his bride, who brought to him a dowry of an equal sum. Eve Esther, a third-generation colonial, was the granddaughter of Abraham Haim DeLucena, who had come to New York at the end of the seventeenth century. It is believed that he was the grandson of the earlier Abraham De-Lucena, who arrived in New Amsterdam from Holland in 1655. If these two DeLucenas were related, then Uriah, a century after the first Jewish settlement in North America, linked the future generations of Hendrickses to a leading figure in colonial American Jewry.

The Gomezes were active in the coastwise trade and in the commerce of the West Indies, and Hendricks' marriage alliance brought him closer to a trade for which he already had a penchant. The independence that characterized his first years in New York had not diminished and, although he did not turn aside any advantage from the Gomezes, he shaped his own business career.

In October, 1763, Eve Esther gave birth to a daughter, Rachel, who was affectionately called Richa. By September, 1764, a second daughter, Rebecca, was born to the Hendrickses, and soon after came a third, Matilda. Uriah Hendricks now concerned himself more with his three little daughters than with the affairs of the colonies.

Instead of the desired prosperity, the peace that was negotiated in 1763 brought with it an intensification of the depression. Hendricks devoted his time to finding an area of business that would be less competitive. The distribution of his imports was now more difficult than before, and the decline in business continued throughout the 1760's. Hendricks' attention was focused on a business that was emerging as one of the colonial necessities. His former Hanover Street neighbors, John Dies, John Troup, Goelet and Curtenius, were ironmongers and metal merchants and had strongly impressed Hendricks with their success. The pig and bar iron that they imported from England was turned into cobblers' and carpenters' tools or agricultural implements as well as nails and stove plates. Records of Hendricks' entry into this new business are meager, but iron and copper goods are referred to frequently in his advertisement at this time. However, it was not until 1764 that ironmongery became a major article of his trade. The prospects of this new venture led Hendricks to consider the possibility of moving to quarters that would be spacious and comfortable for his growing family. The area around the Sloat was a busy place, his stock had again begun to increase, and he felt crowded. Manhattan was already rich in the striking contrasts that characterized its history. If English was the spoken language of the Sloat, the old Slip Market where Uriah first began business was still a street where it was impossible to shop without a knowledge of Dutch. On Mill Street, where the synagogue stood, many of the languages of Europe were spoken with ease, and the Jews themselves used Portuguese, Spanish, French, Dutch, German, Yiddish, Polish, and English. The houses contributed to this multitude of contrasts. At the foot of Broadway stood an impressive group of stately homes—those of Livingston, Van Cortlandt, and Watts. Within a few minutes' walking distance were the wooden tenements on Water Street, with only five or six rooms occupied by two or three families. The houses and buildings below the Fresh Water district were made of roughhewn wood, and unheated quarters were as common in winter as filth and impurity were in the heat of summer. In 1764 a group of merchants, including Uriah Hendricks, urged the legislature to improve the housing conditions in this area. The merchants proposed that all new buildings erected south of the Fresh Water should be built of brick or stone and be roofed with slate or metal. When Uriah finally moved, it was to a residence of such a description near the new Custom House, in Broad Street, one door from the corner of Bayard Street.

Uriah's old business neighbors did not fare well in the peacetime economy. Naphtali Hart Myers returned to London; Hayman Levy, the fur trader and army supplier, was on the verge of bankruptcy; and Manuel Josephson was just ekeing out a living. Privateering had collapsed, the coastwise trade was uncertain, and the English were even more determined to crush the American smuggling trade that deprived them of import duties. The taxes imposed upon the colonies since the end of the French and

Indian War added to this burden. Although Uriah turned more and more to the sale of metal goods he had no intention of discontinuing the other aspects of general merchandising or the sale of lottery tickets, and this was fortunate.

Resentment of the postwar taxes culminated in 1765 in the bitter protest against the Stamp Act. The provisions of the act required the purchase of stamped paper, which was to appear on all documents, newspapers, and almanacs. Funds derived from its sale were to be used to pay for the billeting of the King's soldiers in the colonies. Other provisions of the act were equally resented, and the tradesmen and merchants envisioned a loss of their liberty if the legislation were enacted. Hundreds of New York merchants met at the City Arms Tavern to protest the Stamp Act and agreed not to import any taxed British goods. This compact was so successful that virtually none of the proscribed imports were put ashore in New York. As a result, imports dropped in one twelve-month period from £482,000 to £74,000, and non-importation became the main subject of mercantile conversation.

Although Hendricks was eliminated from active trade when restrictions forced him to retreat from ironmongery, he stuck to the sale of those imports that had not been restricted by the compact of non-importation. Only when the alarm provoked by the Stamp Act had subsided did Hendricks seriously resume the metal business. Accounts were opened with major copper, brass, and iron merchants in London and Liverpool, and the business that was begun in the mid-sixties by Uriah Hendricks was interrupted only by revolution and war.

During these times marked by uncertainty and business irregularity, little Sally, named Jochabed Sarah, was born. A little more than a year later, in the fall of 1768, a fifth daughter, Hannah, arrived. And toward the end of 1769, Mordecai Gomez, Uriah's first son, saw the light of New York. His birth was a joyous occasion for the Hendrickses and Gomezes, and Uriah invited Abraham I. Abrahams to perform the rite of circumcision.

From London, the city of transit for many Jews on their way to America, came Abraham Wagg, a relative of the Franks family and a tradesman with a taste for politics. Shortly after his arrival in New York, Rachel Gomez, Hendricks' youngest sister-in-law, who had just turned twenty-one, found favor with the English bachelor, who was almost twice her age, and the two were married on July 4, 1770. Wagg's brothers-in-law, Moses Gomez and Uriah Hendricks, witnessed the marriage contract, to which Uriah affixed his Hebrew signature. A year later Uriah showed his affection for his new brother-in-law by choosing him as godfather to his second son, Menahem, named Harmon in English, after his recently deceased great-uncle.

All during this busy period Hendricks continued to be occupied with the problems of British imports, and he was frequently engaged in discussions either in his chambers or at one of the business taverns. He strongly favored the resolution to import only articles that were not affected by the oppressive taxation. On July 23, 1770, the resolution was published in the *Gazette,* and Uriah was named among those who were its constant advocates.

About a month after the publication of the new resolution news reached New York that Aaron Hendricks, Uriah's father, had died. Uriah, ordinarily a cheerful person under the most distressing of circumstances, was deeply saddened by the news. He made arrangements to sail for London as soon as possible, but he could not leave until the July packets set sail from New York. Problems had arisen concerning the settlement of his father's estate, which was appraised at £6000. Distrustful of everyone, and still envious of his brother-in-law, Levy Cohen had taken advantage of this opportunity to seize the accounts and records of the elder Hendricks, and he refused to yield them to the estate attorney. The surviving fragments of Uriah's London diary reveal the difficulty with Cohen, but the solution remains obscure. With the aid of Lyon Norden, one of Uriah's London accounts, the matter was somehow resolved. Hendricks returned to New York weary from the harangue with Cohen, saddened by the death of his father, and exhausted by the long voyage home. He found his household of seven small children a busy and happy place.

In the spring of 1772 another son was born to Eve Esther and Uriah, and he was named after his recently deceased grandfather. But the new baby, Aaron, died soon afterward. For the first time grief visited the Hendricks household. Matilda, the sixth daughter, still an infant, died about the same time. The birth of Charlotte helped assuage some of the pain, but a deeper concern, growing instinctively from day to day, had developed about seven-year-old Sally, who responded poorly to the ways and games of her brothers and sisters, kept to herself, and gave signs of being quite different from the other children. With the birth of Esther in June of 1775, the strain proved too much for the mother, who was a frail woman. She died six days later, at the age of thirty-three. "She left a sorrowful Husband and 8 small children, to bewail the irreparable Loss of an affectionate Wife and a tender Mother," lamented the New York press in their obituaries of Eve Esther Hendricks.

Soon after the death of his wife Hendricks moved his business back to Hanover Square, opposite the ironmongery at the Golden Key. One can only speculate about the provisions that he made for the care and education of his children. Of his two sons, Mordecai was lost at sea while returning from London some time after his mother's death. Harmon, the other son, acquired a sound religious education at Shearith Israel. All the daughters received a secular education. Even Sally, who was slower than her sisters, learned to write a fair letter. Deeply involved as he was with the responsibilities of his children's upbringing—the oldest of whom was thirteen when their mother died—Uriah had little time to follow the conflict that was sharpening daily between the British and the colonies.

In the summer of 1776 it became certain that the threatening British fleet would soon occupy New York. An exodus from the city had begun. The majority of the Jews who were able to flee determined to do so, rather than remain under British rule. A number of Jews joined the army of General Washington. Others, the course of whose flight remains unknown, sought safety wherever it could be found. The greatest number eventually found haven in Philadelphia. The minister of Shearith Israel, Gershom Mendez Seixas, because of his known patriotic views, had no choice but to leave. On August 27, 1776, Seixas gathered as many of the ritual treasures of the congregation as he was able and conveyed them with his family to Stratford, Connecticut. The population of New York was greatly reduced, but not all the New Yorkers who were inclined to leave had either the means of going or a safe destination. The number of Jews who remained sufficed for a religious quorum, and the Mill Street synagogue became a silent witness to the prayer for His Majesty, George III, recited by a congregation whose loyalties were decidedly mixed.

In the crucial months that followed, Hendricks was determined not to hazard a move that would affect the care of his motherless children. His ties were English, his family Londoners; but his in-laws, the Gomezes, known as "Staunch friends to freedom," fled the city. Two of the younger Gomezes, Abraham and Moses, brothers-in-law of Uriah, stayed behind and saw the British take possession of New York.

On September 21, less than a month after Reverend Seixas had left for Stratford, a disastrous fire which began in a wooden house near the Whitehall Slip spread rapidly through the older district. It engulfed part of the section that Hendricks and other citizens had petitioned the legislators to improve twelve years earlier. The holocaust made no distinction between the wooden houses of the poor and the stone mansions of the rich. Hendricks' home was untouched. In addition to the problems that the occupying British military had to face in quartering its troops it also had to meet the unexpected circumstance of providing for those left desolate by the destruction of 493 houses. But the Army came first.

Hospitals, barracks, prisons, and lodgings were in immediate demand. First the city's warehouses were requisitioned, then the churches. On William Street the North Dutch Church was turned into a prison; then, one by one, the Presbyterian churches, the Baptist and Friends meeting houses, and the French church were converted to some military use. The Moravians barely escaped having their place of worship turned into a prison. Only St. Paul's and St. George's the loyal Anglican churches, were left untouched by the British. For some reason the synagogue, too, was spared.

Entrenched securely in New York for the next seven years the British turned the city into a center of civil and military activity. Wherever they could be detected, the remaining Whigs were routed out. The colonial tradesmen who remained behind, for whatever reason, had no choice but to support the English military government and publicly acclaim their sympathies for the King. In October, before the debris of the great fire was cleared away, a meeting of the city's leading men among the merchants, the clergy, the gentry, an overwhelming number of loyalists, and those who had to remain publicly asserted their fidelity to "his Majesty's paternal goodness." Of the total of 948 signers of this document fifteen were known Jews, and of these only one, Barrak Hays, proclaimed himself a loyalist. Two of the synagogue functionaries, Abraham I. Abrahams and Levy Israel, entered their names in the declaration along with Uriah Hendricks and the Gomez brothers. Moses Gomez remained in New York long enough to see the family fortune confiscated by the Tories; then he slipped through the lines to Philadelphia. Here he joined the other exiled Gomezes and those Philadelphians, like Jonas Phillips, who earlier had been trapped by a similar dilemma. Phillips, a soldier of the Pennsylvania militia, lived through the British occupation of Philadelphia with a family of seventeen children. Uriah Hendricks ventured to do the same in New York with his family.

A second loyalist declaration was presented to the British on November 28, 1776. Uriah's English brother-in-law, Abraham Wagg, the wholesale grocer, who had not joined in signing the first declaration, subsequently announced his loyalist views. Wagg, distressed by the conflict, later submitted a plan of reconciliation for the colonies and the mother country which went unheeded. Uriah Hendricks' own record during the Revolution, but for a few scattered references, is vague. Toward the end of 1777 and very early in 1778 he offered for sale "rice in Tierces." Thereafter his old trade dwindled away and only the less lucrative business of a wartime lottery office remained on Hanover Square. The tempting trade of coastwise privateering, at which he was an old hand, lost its lure; he was not to be found at the daily auctions of merchandise taken from captured ships, and the vast imports that poured in from London were not consigned to him. It may have been with a feeling of nostalgia that he read in James Rivington's *Royal Gazette* of his former account, Moses Franks, who was raising a fund of money in London for those affected by the war. It was the only "Synagogue Intelligence" that appeared in the Tory press in seven years, except for the one notable exception.

On the night of Tuesday, September 10, 1782, an unknown number of British soldiers forced their way into the Mill Street Synagogue and removed the Holy Scrolls, one of which had been deposited by Uriah Hendricks. Barrak Hays, who appears to have assumed authority over the synagogue, politely offered five guineas' reward for their return:

. . . any Person who will deliver to Mr. Barak Hayes, at No. 56 Smith Street, two Setts of Parchment Rolls, written in Hebrew Characters, and ornamented with a Crimson Damask, trimmed with Gold Lace and Fringe, taken out of the Synagogue last night. If delivered with or without the Drapery, will be entitled to the reward, and no questions asked. New York, 11th September, 1782.

The Scrolls were recovered but found to be irreparably damaged, and two soldiers involved in the desecration were caught and flogged by the military authorities.

The number of Jews in occupied New York was greater than is generally believed. Lyon Jonas of London continued his trade as a furrier; Levy Simons as an embroiderer "to the gentlemen of the Army"; Moses Hart sold his bottled porter and hams; Samuel Lazarus distributed dry goods, Benjamin Raphael was a fruiterer, Joseph Abrahams a tobacconist and distiller, and Samuel Levy was an English tanner. Lyon Hart's company offered silverware and jewelry. To help entertain the officers of the Army and Navy, Isaac Levy of London exhibited his "Dexterity of Hand" at Roubalet's Tavern, a favorite spot of entertainment for the British.

At most there were thirty Jews and their families in New York City during the occupation. In 1779 there was a notable addition to this small community. He was Alexander Zuntz, a Hessian provisioner who traveled with Howe's army. He remained in Philadelphia during the nine-month occupation of the city and then went with the Army to New York, where he actively engaged in business in Queen Street, near Beekman. His goods came in from London with the fleet, and he could supply a wide variety of items from smoked oxen tongues to a "Steiner violin" for connoisseurs. Zuntz had no prior ties with the colonies and owed his allegiance only to the Army, with which he came.

Late in 1782 the Jewish community began to decline. The *shamash* Levy Israel died; Abraham Wagg returned to London; and Barrak Hays, since the fortunes of war now favored the colonial armies of Washington, informed the public that he was bound for Europe. But then Hays decided to follow the advice of the *Royal Gazette,* which recommended the fine climate and choice opportunity for business in Nova Scotia. He fled with the Tories who went north. Following Hays's public announcement of his departure Samuel Lazarus, Samuel Levy, and Alexander Zuntz made known their intentions of returning to London. After Hays departed, however, Zuntz, unexpectedly assumed temporary responsibility for the synagogue; and Lazarus and Levy, reversing their plans, also remained in New York.

When victory for the colonists was certain, the Tories, and then the troops, gradually departed from New York. By the time that Washington's army reentered the former stronghold on November 25, 1783, it was no longer a hostile loyalist center. The Jews, the largest number of whom had found haven in Philadelphia, began to return. Hayman Levy, Isaac Moses, the Gomezes, and the other patriotic New Yorkers came back to their homes.

Within weeks of their return the former exiles seated themselves in the vestry room of Shearith Israel and made plans to reorganize the congregation. At their first meeting Alexander Zuntz resigned the office of president, which he had filled until the British evacuated the city. Hendricks, with five other residents of wartime New York, reemerged from obscurity to join the homecoming refugees in drafting an address to Governor George Clinton on the success of the colonists. It was debated whether or not the address should be presented in the name of the former exiles alone, or in the name of the congregation. It was a tribute to Hendricks, as a resident of Tory New York, that his motion, made jointly with his patriot friend, the goldsmith Myer Myers, was adopted, and the address was sent in the name of the congregation. Other matters vital to Shearith Israel were given immediate attention. Hendricks and Myers were elected to superintend the cemetery, which had been neglected during the war; the synagogue was in need of restoration; and Seixas, the minister, was recalled from Philadelphia, where he had assumed authority over the religious affairs of Congregation Mikveh Israel, the oldest in that city.

2

HARMON HENDRICKS
The Making of a
Merchant

I

At the time of the reorganization of the synagogue Uriah Hendricks' daughter Richa had reached the age of twenty and Esther,
his youngest daughter, had just turned eight. His only surviving
son, Harmon, celebrated his bar-mitzvah in the spring of 1784
and a year later, at the age of fourteen, he began to work for his
father. He learned to clerk, write the duplicate and triplicate
letters that were sent by the packets, keep accounts, and count
pounds in the same fashion as his father had before him. It was
no idle boast for Harmon to write later that he was rarely guilty
of an error in reckoning.

The Hendricks business expanded rapidly. Ironmongery, cutlery, and copper imports were basic commodities of the shipping
trade when business was resumed with England during the mid-
1780's. Harmon took to it naturally, having inherited his father's
taste for business and his love for the trade across the seas. New
accounts were opened in London, Hamburg, and Amsterdam.
The major English exporters of copper, lead, and spelter, Pieschell
and Brogden, and John Freeman and Company, once again
looked to Hendricks' orders. To these items were added pig and
bar iron, sheet copper and copper nails, and large quantities of

leather and dry goods. Trade was opened with Paris & Company of St. Petersburg, Russia, and resumed in the West Indies with Isaac Gabay, David P. Mendez, the De Leon family, and a host of other merchants whose names filled the huge Hendricks ledgers.

Synagogue activity, which became a vital part of the life of every generation of the Hendricks family, was an intimate concern of Hendricks. Characteristic of this interest were his regular contributions to charity funds, aid toward enlarging the cemetery, and support of the declining Newport synagogue, to which Hendricks was one of the first contributors. His philanthropic interests were not limited to Jewish life. He was an annual subscriber to the city dispensary, which provided free medical care, he contributed liberally to the French refugee fund, and he personally aided his less fortunate business associates when they were in economic distress or in ill health. He established a pattern of charitable conduct that was later emulated by his son Harmon.

Business connections in Rhode Island and interest in the Newport synagogue brought Hendricks into friendly contact with the children of the deceased Newport shipping merchant Aaron Lopez. Hendricks' brother-in-law, Moses Gomez, had married Esther Lopez, and it was Esther's younger sister whom Uriah Hendricks sought for a second wife. In 1787, twelve years after the death of Eve Esther, Uriah Hendricks married seventeen-year-old Rebecca Lopez. The Hendricks' household now moved to 112 Pearl Street, where the sons of Hendricks' colonial colleagues eyed his marriageable daughters as they came of age. In 1789 Hayman Levy's son, Solomon, married Rebecca, and that same year the West Indian Jacob Cohen De Leon, the forerunner of an important South Carolina family, married Hannah. Richa, the oldest daughter, married Abraham Gomez, Jr., of Bordeaux, France. Abraham was not related to the New York Gomezes, one of whom, Benjamin, a publisher and bookseller, married Charlotte Hendricks. Only Esther and Sally remained unmarried during their father's lifetime.

On one of his Newport trips, late in 1789, Uriah spent the Sabbath weekend in the city, but in protest against the manner in which the synagogue service was conducted he deliberately abstained from attending public worship there. The reader of the service used a printed Hebrew text of the Pentateuch, although law and usage required that the reading should be from a parchment scroll. Hendricks' protest to Moses Seixas of Newport about this practice resulted in a rabbinic query by Seixas to Manuel Josephson. Josephson, a linguist and rabbinic scholar, advised Seixas on the ritual impropriety of reading from a printed text and supported the view of his friend Uriah Hendricks.

Six months later Uriah, as one of the *Adjunta,* or managers of Shearith Israel, addressed Moses Seixas in the name of the congregation on a different matter of importance. The *Adjunta* requested the cooperation of the Newporters to join with the other Jewish congregations in a single address to George Washington to congratulate him upon his election as the first President of the United States.

In 1791 Hendricks was elected president of Shearith Israel, and his interest in all matters concerning Jewish welfare motivated him to sign a letter of introduction for two Polish rabbis soliciting funds on behalf of the Jews of Palestine. The bearers of the letter traveled as far south as Charleston, South Carolina, in the fall of 1791, where they presented their plea for funds to Gershon Cohen, a soldier of the Revolution and the city's rabbinical leader. The Jews of Charleston had just acquired a synagogue site for £400, and their depleted resources prevented them from helping the two Polish messengers. In explaining this situation Cohen described a Sunday religious service held by the congregation in order to raise funds for the local orphan school. Hendricks must have communicated the fact that he was appalled by the news of a Jewish religious service on Sunday, for he received another letter from Charleston soon afterward explaining that nothing had been done in violation of Jewish practice. When funds could not be raised from the general public on the Sabbath, the meeting, with a brief prayer, was deferred until the following Sunday.

It was not unusual to find Uriah Hendricks at the Tontine Coffee-House in Wall Street, New York's bourse, accompanied by young Harmon. Here they might chat about prices current, or nod to a Livingston, a Varick, a Brinckerhoff or to Dr. Samuel Bard, men with whom they were acquainted personally and professionally. From Wall Street Uriah might hurry along Broad Street over the familiar route that he had walked for almost forty years. The population of the city had almost tripled since his arrival in 1755, and it was increasing rapidly, with more and more of the newcomers crowding into the tip of Manhattan. In Broad Street he stopped at John Slidell's soap and tallow chandlery to change his usual order.

Slidell's origin was a mystery. He was purported to be a Jew, but actually he observed no religion at all. He was an ardent patriot, but refused to budge from the city during the British occupation. On that day in 1792, the last meeting between the two men—for Slidell died soon after—both would have been astounded to learn that seven decades later their descendants were destined to meet in official capacity as representatives of the Confederate States of America: Edwin De Leon from Charleston and John Slidell from New Orleans.

Jacob Cohen De Leon, the husband of Hannah, and the oldest of Uriah Hendricks' sons-in-law, was somewhat of a rolling stone. He entered business in Philadelphia, where his first son was born, but a few months later moved to New York, where his best customer was Uriah Hendricks. Neither city seemed to suit his needs or enhance his abilities, and off he went with his family to Kingston, Jamaica. The venture to the islands was as brief as it was impractical. Hannah became ill, business was poor, Jamaica was about to erupt in a series of slave revolts, and there was little of promise that remained on the island for the De Leons.

Occasionally, to assuage her father's concern, Hannah De Leon sent him a gift of marmalade, and her letters were full of pleasant gossip about the Jewish island peddlers. Uriah must have chuckled with gratification as he read about Joe Correa, who "run himself out of breath for the space of three miles" in order to join his family without breaking the Sabbath.

The De Leons returned to New York, where Uriah Hendricks again became Jacob's best customer. In 1796 Jacob made a second trip to Kingston, shipped some merchandise to his brother-in-law, Harmon, and finally, in 1799, settled in Charleston in the hope of benefiting from its new economic upsurge.

Solomon Levy, Rebecca's husband, was as settled as De Leon

was restless. In the early 1790's, shortly after their marriage, the Levys moved to Peekskill, New York. Levy was an ambitious shopkeeper who worked twice as hard as his competitors to earn the same amount. He was rich in dreams, but did not possess the venturesome nature of his father, the colonial trader.

Little is known of the French brother-in-law, Abraham Gomez, who took Richa for his bride. After a brief residence in New York he returned with his wife to his native Bordeaux.

Charlotte's husband, Benjamin Gomez, was related to Abraham only in name and as a brother-in-law. The record of his career is livelier and more complete than that of the French Gomez. Benjamin was a publisher, bookseller, and stationer, and his shop in Maiden Lane, where he also sold lottery tickets, gave every appearance of success. To Gomez belongs the distinction of being the first Jewish publisher in the United States. His book trade thrived sufficiently to encourage Naphtali Judah, a future Hendricks brother-in-law, also to open a book and lottery office.

Lotteries provided the legal means of raising the funds that paid for building many of the churches, colleges, and public institutions of early America. It was a business that Uriah Hendricks never let slip out of his hands. Both Gomez and Judah often visited the Hendricks' lottery office, where they could purchase sheets of uncut tickets for resale. In one week alone Naphtali Judah paid for $550 worth of tickets and his brother Aaron settled for a similar amount in English pounds. It was a rewarding bonus for a going business, but it represented only a segment of the economy of the time. The close contact between Gomez and Judah that brought them together in the book trade in 1795 later united them as partners in a lottery office. Benjamin Gomez dealt cautiously, but Judah, eager for grand returns, was dazzled by the wheel of fortune.

Hendricks was proud of his sons-in-law, all of whom strove for success during the first decade of the new America. The old frontiers that had challenged their fathers—the Indians and the fur trade—had lost their lure. Gomez, the intellectual, was content with a book and lottery office; De Leon, the petty tradesman, longed for business stability; and Levy, the settled storekeeper, continued his hunt for small but profitable bargains. Harmon was the only member of the family who, upon reaching maturity, had the ambition, the determination, and the zeal to pioneer on the frontiers of industrial America.

II

Uriah Hendricks' Pearl Street home was located in the middle of a diversified setting, not far from the Battery. Close by, Edward Livingston, the distinguished lawyer who later became Secretary of State, acquired two houses with surrounding plots of ground. Within a few hundred feet Archibald Gracie, the merchant-banker, chose the site for his stately mansion. On the southern corner of Pearl Street, fronting the Battery, Gardiner Baker opened a menagerie in 1794. Here he exhibited a variety of wild animals, which could be seen for a small admission fee. Squeezed between these buildings were a few printing shops and a number of small homes.

Under the guidance of his father, Harmon Hendricks was sent on a tour of the states to become acquainted with the American market. Shortly after his return he embarked on a voyage to the West Indies to learn the island trade and meet the men with whom he was to engage in business. He visited Jacob De Leon in Kingston, spent some time in Spanish Town, and made friends with Solomon Flamengo, Isaac Gabay, and David P. Mendez. He returned to Pearl Street in 1792 after an absence of almost a year. His apprenticeship completed, he was eager to venture into the business world on his own. With money that he had saved and the assurance of a loan from his father, he rented a storehouse in the Old Slip. Harmon Hendricks was determined to enter the island trade in preference to the carrying trade with Europe.

At the age of twenty-one Harmon was thoroughly familiar with the ships, their captains, and the manner of maritime commerce. Excitement spread along the docks when the first March packets and the spring vessels sailed into New York harbor after a winter's absence. A Liverpool bark bringing Harmon Hendricks a parcel of 32,700 "smoking segars" from Kingston, or a Dutch merchantman carrying a consignment of 300 pounds of oil cloth from Amsterdam, or a trim New England sailing vessel stopping to discharge a load of indigo from Charleston signaled the opening of the spring trade. Harmon met the captains, checked the bills-of-lading, and arranged with the draymen to haul his goods through the jostling waterfront traffic to his storehouse in the Old Slip. With the arrival of each shipment he yearned for the day when he would have a vessel of his own and a sea captain at his command. Each cargo, each invoice, and each transaction —from the sale of two lottery tickets for Harvard College to the importation of 4,000 pounds of Spanish coffee—was a challenge.

By the time Benjamin Gomez had decided that the sale of lottery tickets was an important adjunct to his book trade Harmon Hendricks was completely absorbed by the West Indian trade. He found European commerce, to which he was trained, much less appealing. Contacts with the island merchants, made through his brother-in-law De Leon, offered young Hendricks the opportunity to exploit his knowledge of the export and import trade. His adventure to Jamaica with a vessel of American goods consigned to the merchants of Kingston and Spanish Town was profitable, and other shipments soon followed. Hickory canes, hair ribbons, umbrellas, watches, cheap jewelry, and barrels of New England onions, flour, or apples were shipped regularly from New York to Jamaica. In return the Jewish island merchants, Solomon Flamengo, David Pollander, Joseph Correa, David P. Mendez, and the Gabays dispatched to him such produce as coffee, quince pears, pimento, Jamaica ginger, "Spanish segars," and puncheons of castor oil. Harmon soon discovered that Cayenne pepper was a trifling article and unprofitable. Rum enjoyed better sales; cigars, when properly packed to retain their flavor, had a ready market; and "Castor oil and old metal bring a good profit."

When the Jamaica merchants visited New York in 1795 they came to the Hendricks household, and Harmon joined them in the discussions of the new commercial treaty between America and England. He wrote that, "the difficulty of obtaining freight . . . will induce me this summer to purchase a small vessel as soon as the negotiation of Jay with Britain is published & we have no doubt will be much to our advantage in trading with the Islands. . . ."

Harmon had a zeal for work and he would occasionally slip away from the company to another quarter of his father's spacious house to finish the day's correspondence. If news reached him that favorable winds permitted one of the packets to sail he would hurriedly deliver his letters for the West Indies to the ship's captain. He admitted to Solomon Flamengo, for whom he described his conduct, that "the jolity & agreeable songs rings reproaches in my ears for my absence" from the company.

Gradually Harmon applied the astute caution emphasized by his father in all matters of business. He studied the sea lanes, subscribed to the latest edition of Pinkerton's *Geography,* avoided the illicit island trade, and carefully read the speeches in Congress about privateering or the new protective tariff law. In 1794 he became a member of the Tontine Coffee-House, entered his name with the United States Office of Discount and Deposit, opened accounts with two New York banks, and affiliated himself with the institutions necessary for the success of an enterprising young merchant.

Harmon had witnessed the resumption of trade between his father and the English merchants after the Revolution. The elder Hendricks was primarily an ironmonger who also imported vast quantities of copper, tin, lead, pig, and bar iron from Bristol and London. In the course of the sale and distribution of these metals it became apparent that the demand for them was far in excess of what the Americans could produce. The American consumer after the Revolution was as dependent upon manufactured English goods as he had been when he was still a colonial subject. Americans still brewed English tea in the brass or copper kettles that were made in Bristol; their food was still served on the shiny tin plates that were cut and edged in the London mills; and those who wore brass buttons on their waistcoats knew that the foundries of Great Britain had produced them. The soap boiler who required thick sheets of copper for his vats, the small farmer who converted part of his corn and rye crop into whiskey and needed good copper bottoms for his stills, and the coppersmith who used a variety of bars, rods, and plates, all could turn to Uriah Hendricks, satisfied that their needs would be supplied. New York City's brass founders, coppersmiths, tin stores, plumbers, and pewterers were the backbone of Hendricks' local trade. But there were other customers in the inland sections of the state and large consumers in the bordering states of Connecticut and New Jersey. Specialists in the trade, such as Samuel and Judah Myers who turned copper into decorative artifacts, or Jacob Marks, who urged the first Mint of the United States to buy his refined copper, were Hendricks' regular accounts at the time when Alexander Hamilton's "Report on Manufactures," which proposed a system of tariffs for industry and various internal improvements, and encouraged domestic manufacture, went unheeded by those whom it affected the most.

Copper sheathing was an accepted part of English shipbuilding after the American Revolution. When sheathing was fastened to the submerged portion of a ship's hull it prevented wood rot and worminess. Copper nails and spikes replaced those made of cast iron and were used to secure the seams of sheathing. The nails penetrated the toughest oak and fir bottoms and blended smoothly with the surface of the sheathing. This technique repelled barnacles from clustering to the seams and spreading along the ship's bottom. Such advantages liberated a ship of war from impediments that reduced her speed; enabled her to be repaired more quickly; and permitted her to stay longer at sea. Vessels that plowed the tropical waters of the West Indies, or made the long voyage to the China seas, or went on sealing and whaling expeditions were especially vulnerable to the deterioration that could be reduced or prevented by the use of copper sheathing.

In addition to the metal trade Hendricks conducted a large business in fine English leathers, choice Dutch linens, and woolens made in Leeds. Huge bales of white duck and diaper cloth, known as Russia goods, were shipped to New York from St. Petersburg by way of London. In addition to these major importations there was an interest in minor enterprises, such as obtaining the proper ingredients for the manufacture of *kasher* cheese. The steady rise of importations into the United States at this time was astounding. From $29 million in 1790, importations soared to $111 million in 1801. Uriah Hendricks was an integral part of the transatlantic commerce that challenged the English hegemony in the first years of American independence.

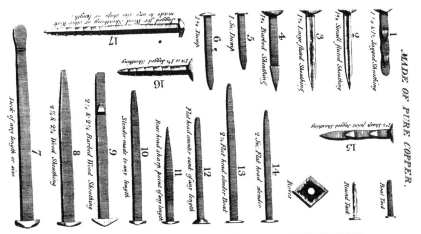

MANUFACTURED BY S. GUPPY *(PATENTEE)* BRISTOL.

18

Improved Deck Jagged or Plain of any size or length.

BRISTOL,   1806.

I BEG to recommend to your notice and use my PATENT PURE COPPER SHEATHING NAILS (with two or three barbs or jaggs) counter-sunk heads, so formed as to hold much faster than a taper cast Nail. They are so well hammer-hardened, and the pores of the Copper so closed, that they can be driven into the hardest wood, or a ship of the greatest tonnage sheathed with them, without one in a thousand bending, breaking, or the heads falling off by driving. The certainty of the jagged Nails holding, and fastening the Copper on the Vessel's Bottom, be it of soft or hard wood, old or new, has been satisfactorily proved in various instances. The jaggs or barbs hold as they drive, even if the Copper is 28, 30, or 32 oz., and neither tear the sheathing or ship's bottom to pieces, by requiring such enormous large holes as are generally made for cast Nails.

From the quality of the Patent Nails, Sheathing can be nailed on as tight as the Builder pleases. No one need be told, the closer the Copper is fastened to the bottom the better—that a *smooth surface* (which the Patent Nails leave) will last twice as long, and a Ship sail much faster, than with a rough bottom, and uneven surface ; and it is impossible to fasten the Copper close with cast Nails, for if they are driven up, the heads of half will fly off, in consequence of the brittle nature of the metal ; the head not being close will impede the sailing, catch grass, weeds, &c. Those driven into a hole, where a Nail has been broken, cannot be driven up, whereas the Patent Copper Nail will admit of being driven with the Copper into the wood ; it cannot be easily broken, and dents in smooth with the sheathing.

The injury done to Ships' Bottoms, as well as the Copper, by the use of large cast Nails, *has been a subject of great complaint ; and barnacles are frequently found on the heads of cast Nails, which very much impede the Ship's sailing.* In Oak or Fir Bottoms they exceed any thing ever yet used, and are a saving of full 50 per cent. besides such as are taken off a Ship's Bottom, are in value equal to any old sheets ; whereas the mixed metal are not above two-thirds the value, and, if melted with old sheathing, reduce its value and utility. There cannot be a doubt of the ship's sailing much faster, as well as the Copper lasting a considerable time longer, with the Patent Nails. The many complaints of bad Copper has been fully ascertained to arise from the creases or wrinkles between the Nails, the projected Sheathing rubs off, and leaves the Copper in holes. Were the Patent Nails to cost twice as much per lb. as the cast mixed metal, the saving would be considerable.

The Ship Builders and Owners, who have used the Patent Nails, have declared, and convinced me by a repetition of their orders, that they will never use another nail, when they can be supplied with Patent Nails.

PURE COPPER NAILS for WOOD SHEATHING, on an entire new plan, which do not in the least injure the wood, as large cast Nails do (being jagged, and about half the size of cast Nails, the wood sheathing is not liable to fall off) and the advantage in favor of the Ship-Owner is equal to the Copper Sheathing Nails.

If 70 Nails are actually used in a sheet, 3,143 sheets (the quantity used for a 74-gun ship) will want only 220,010 Nails ; but the charge is cwt.26 : 3 : 24, which, at 85 in the lb. amounts to 246,700 Nails ; so there appears to be 36,690 nails charged more than are actually used ; yet this is by no means an improbable quantity broken.

In the East-India ships cast Nails considerably heavier are used ; of course the difference is much more. About 160 are sufficiently large for any Vessel ; 200 to a pound will be found large enough for Vessels of 200 tons.

☞ Be careful to use a Punch not larger than the Nail and not above ⅞ of an inch long.

SAMUEL GUPPY, Patentee, Bristol, or Dowgate Wharf, London.

A rare and important description of the copper nails used in ship sheathing, 1806.

The English metal merchants who vied for his trade—John Freeman and Company, Stratton and Gibson, Pieschell and Brogden, Thomas Holmes and the Bolderos—were also introduced to Harmon. Trade established decades earlier by Uriah Hendricks with his brothers-in-law, the Cohens, the Oppenheims, and the Pollocks of London, was now continued by their children. An import business was also carried on among these branches of the family that totaled thousands of pounds annually and kept Harmon alert to English commerce. The most active of the London cousins were the Oppenheim brothers, who supplied everything from pins to pianofortes. To this group of English relatives-suppliers, Harmon added a fourth account, Mordecai Gomez Wagg, his mother's nephew. He traded with Wagg independently of his father, and while his volume of business in England was insignificant by comparison, it enabled Harmon to obtain many of the items necessary for reshipment to his customers in the West Indies.

Early in 1797 there was talk of an embargo and a fear of war with the French, whose navy was choking the shipping trade by preying on English vessels bound for American ports. Harmon cautiously advised his favorite customers of the American reaction to these unexpected events. To Daniel Crommelin of Amsterdam he sent advice not to ship by way of Jamaica, for "almost every vessel from and to British Ports are taken and condemned by the French." Jamaica merchants, many of whose offices were in New York, were unable to sustain these losses and were pushed to the verge of bankruptcy. Harmon confidentially forwarded information about the financially imperiled merchants to David P. Mendez of Kingston, who in turn exchanged valuable information on prices-current with Harmon. French depredations of British shipping opened new opportunities for the entrepreneur. "The door of speculation will again be opened at your place," wrote Harmon to Mendez, but neither one had the opportunity to speculate by using American instead of British ships to elude the French privateers.

The final distributors of Harmon's West Indian goods were the domestic shopkeepers. Much of his merchandise was reshipped to David Lopez in Charleston, or to Jacob Levy in Wilmington,

North Carolina. Some went inland to Myndert Lancing in Albany, and a considerable amount went north to Hartford, to James Ward, customers to whom Harmon was introduced by his father. Huge consignments of cigars were sent to nearby Philadelphia to be sold at auction by Sol Marks or peddled in the center city taverns by Abraham Cohen. Harmon had not yet sold any of his father's more affluent customers, such as Roosevelt, Livingston, or Brinckerhoff, but depended considerably on local shopkeepers and the nearby country trade. He became a regular supplier of goods to the brothers-in-law Gomez and occasionally to Solomon Levy in Peekskill, New York. Lottery tickets were an additional source of income to Harmon, and most of the tickets that reached Jamaica from New York passed through his hands. He became so enamored of the island trade that he confided to Mendez his desire to settle in Jamaica for a period of four to five years. Mendez enjoyed Harmon's confidence at a time when Harmon was writing with undisguised restraint to his other associates: "I am a man of so few words."

The memory of his visit to Kingston and his successful business ventures in Jamaica may have encouraged Harmon to think of settling there. He was aware of the hazards and vagaries of the island trade and had not forgotten Jacob De Leon's lack of success. His own good fortune in New York did not satisfy him. Harmon was eager to expand his business into other fields. He spelled out the terms of this interest to his island friend, David P. Mendez:

Harnassed as I have been diurnally in the reins of a regular trade for more than 13 years is sufficient proof that I ought to posess a general knowledge of business allowing a man common abilities. This advantage I should wish to avail myself of in a Partner therefore I should have no objections to enter with any young man, of good connections and not habitually a game-ster. . . . my own interest will dictate me to be punctual myself and attend particularly to it allowing my partner as capable as myself, but with one of the above description. I would only enter in consideration of his having 2 dollars to one of mine in stock on equal proffits, and after 3 years to with-draw the excess unless I agreed to make good my deficiency to the original stock or dissolve. I trust if such a person should come under your observa-tion you will think of me . . .

By the time Harmon's letter reached Mendez, the yellow fever, which plagued America annually, had once again blighted New York and Philadelphia. Concern for business declined as the number of fever victims increased and when Harmon wrote to Mendez again it was in a different mood:

Philadelphia is again visited with the Yellow Fever which carries off 60 a day on an average of the remaining inhabitants, ⅚ of them having removed in the Country or encamped in the Suburbs . . . New York is again attacked with an epidemic that carries off 30 a day. I have lost many valuable acquaintances, the contagion does not equal Philadelphia and we may expect health again restored a month sooner than our Sister state . . .

Optimism about the restoration of health in New York was deceptive. Before Harmon's letter left the city on the Brig *Chatam* a young Jewish medical student, Walter Judah, who had attended the victims of the disease, fell victim to it himself. When the fever subsided in the fall of 1798 it had taken a toll of more than two thousand lives. Judah was one of the ten known Jews who succumbed. The New Yorkers were as baffled by the disease as the Philadelphians, and the medical profession held differing opinions about the causes of its origin and the procedures for its treatment. In one of his letters Harmon linked the start of the disease to the suspension of trade with the French islands, a large consumer of American beef. "Many stores & Cellars containing from 500 to 5000 Barrels Beef for want of a Market Putrefied and rendered the air contageous in the environs of those Stores, the people thereabout fell the first Victims of it, and since the Beef has been thrown in the North River."

Those who could flee took refuge in the suburban village of Greenwich, a few miles to the north. The height of the epidemic and the unseasonable warmth of mid-September coincided with the Jewish holy days. On the second day of Rosh Hashana, the Jewish New Year, Uriah Hendricks was attacked by the fever. Two of the city's outstanding physicians, David Hosack and Samuel Bard, attended Uriah at his Pearl Street home. In a few days he improved considerably and was able to attend synagogue on Yom Kippur and fast on that holiest day of the Jewish year. As soon as the holiday was over the Hendricks family moved to

Greenwich. Harmon described the subsequent events to his relatives in London:

> . . . no inconvenience appeared to be the result and next day we moved to this place as the town was nearly evacuated, on acc.^t of sickness. The change of air seemed beneficial to his health 'till the morning of 17 Tisri 27 Septr when on visiting in his apartment at 8 o Clock I observed his countenace florit he assured me he felt well, a moment after he leaned on one side, expired without a groan, struggle or sigh, surely heaven in its dispensations of blessings to the best of men could not with more comfort to himself have rec^d his soul.

The members of the Jewish burial society, in fear of the fever, had joined in the flight from the city, making it almost impossible to conduct a proper burial service. Had it not been for the young Jewish musician Samuel Hart, who bravely returned to the city to solicit the aid of Menahem Isaacs to dig the grave and make the necessary arrangements, Harmon's problems would have been multiplied. In the middle days of Tabernacles Uriah Hendricks, "with scarce a gray hair, nor wrinkle, without ever using spectacles, with a heart usually light and gay as a youth, esteemed by every one and his word of equal security to those who held it as his bond," was laid to rest in the Chatham Square cemetery at the age of sixty-one.

It was with considerable emotion that Harmon, not yet thirty, wrote the news of his father's death to each of his English relatives. "To me is reserved the most mournful, the most distressing of all tasks, that of announcing to you the death of my beloved, respectable father . . ." Even at this sad juncture Harmon refrained from repeating the same words in his letters to other relatives, to friends, or to his father's business associates. His imaginative, picturesque, and graphic English told the same story, but always in different language. The man who was soon to enter into a major import business that depended heavily for its success upon the effectiveness and clarity of a letter was a master of epistolary style.

When the thirty days of mourning had passed and the crisp November days of 1798 finally set to rest the mosquito that carried the fever, Harmon returned to New York to begin the settlement of his father's affairs. Uriah died intestate, and Harmon was appointed administrator of the estate. Hetty, as Esther was called, and Sally, who were both unmarried, now became their brother's responsibility. Uriah Hendricks' interests abroad were scattered throughout England and Holland and extended to St. Petersburg, where he kept a fund on deposit. At this point it was learned that the will of grandfather Aaron Hendricks, who had died in London in 1771, included provisions for his American grandchildren that became effective upon the death of Uriah.

Overnight Harmon Hendricks determined the steps he would take. He first declared his plans to the copper merchants John Freeman and Company of Bristol. "I intend to give up the West India trade to follow the track in which I served my apprenticeship with my late father. I solicit your friendship towards me on the same terms you were want to execute the orders of your late friend . . . I being an only son & have married and unmarried sisters & step mother without children." Harmon immediately placed a huge order with Thomas Holmes for tin plates, iron wire, block tin, spelter, bar lead, and brass. Stratton & Gibson were advised that Harmon was the only male descendent carrying the Hendricks name and that the business was to be continued. Pieschell and Brogden recalled Harmon as the confidential clerk of his father for eight years. One by one, all of the English merchants and metal houses were advised of Harmon's intentions and each, without question, extended the same credit and shipping terms that had been allowed to the elder Hendricks. The goods that arrived during the yellow fever for Uriah Hendricks were removed from the glutted public storehouses and prepared for sale. A large stock of leather skins, bales of Russia goods, blocks of tin, nests of brass kettles, layers of copper sheeting, and nails of every description had to be liquidated. It was from this stock that the United States government made its first major purchase of copper, consisting of 485 sheets weighing 5,680 pounds. With the exception of his friend David P. Mendez, Harmon planned to discontinue trade with the West Indies as soon as it was practical.

Harmon threw his energies into settling his father's estate within the time allotted by law. His brothers-in-law showed full confidence in his ability and judgment, but plagued him with dozens of requests and favors. Wherever possible Harmon granted their wishes. Benjamin Gomez preferred a share of the fine leathers to bind his books; Solomon Levy demanded goods for his country store in lieu of cash, but then compromised on sufficient funds to purchase larger quarters in Peekskill; and Abraham Gomez bought dry goods at cost rather than permit their resale to others for a higher price. Jacob De Leon, who had just returned to Charleston, was worried about his share of the estate until he realized that Harmon was a faithful friend who treated everyone fairly. Provisions were made for the widow Hendricks, who returned to Newport to live with the Lopezes, and the court appointed Harmon as guardian for his sister Sally.

While Harmon was occupied in settling his father's estate he met Frances, the daughter of Joshua Isaacs. He courted her for one year and in June, 1800, they were married. Hetty, the youngest of the Hendricks daughters, and Sally continued to live with their brother in the Pearl Street home. There Naphtali Judah, an eligible bachelor who made no secret of his intentions, was a daily visitor. From Harmon Judah obtained considerable credit, cash without interest, the security for a house, and an interest in a paper mill. Harmon's shrewdness in business was not a match for Judah's cleverness in deception. Judah's fortunes took a leeward turn, and "this rascal about a month before he failed, obtained a promise of marriage from Hetty." Naphtali Judah went bankrupt under the new insolvency act, and involved his brothers and both the New York and London branches of the Hendricks family. As the authoritative head of the household Harmon arranged a conference of his brothers-in-law in the hope of persuading Hetty not to marry Judah, but to no avail. Hetty, now in her middle twenties, feared that she would remain single if she allowed Judah to slip away. By this time Harmon had nothing but contempt for Judah, "the man of honor as he stiles himself whose love for Hetty" might disgrace the Hendricks family. Hetty kept her promise to Judah and married him in 1801. Meanwhile the house at 112 Pearl Street was sold, another was purchased close by, and Uriah Hendricks' estate of $56,000 was settled.

3

COPPER CROSSES
THE ATLANTIC

I

From his new quarters at 120 Pearl Street Harmon Hendricks reentered the general carrying trade. The port of New York, with which he was intimately familiar, was busier than ever. American ships cleared for all parts of the world. Clippers set sail for the China seas with local goods and returned with rich cargoes of silk and tea. Vessels laden with pot- and pearl ash, pitch and turpentine, embarked for English ports where they were reloaded with pig tin, brass kettles, and sheet copper. The produce trade with the West Indies, resumed after the American Revolution, had grown into a lucrative business. The New York merchants benefited accordingly from this overseas commerce.

The upsurge in American foreign trade was a result of the outbreak of war on the European continent, which had disrupted the normal flow of shipping between Europe and her colonies since the late 1790's. England suffered considerably when her maritime strength was deployed for purposes of war, and the Americans, recognizing a grand opportunity, built their shipping into a vast enterprise.

The war that made shipping profitable for the Americans also added to its hazards; cargoes were seized by the belligerent powers, and vessels were lost and destroyed, but the ultimate returns made the risk across the sea worthwhile. The shipbuilding industry, directly affected by the rise in trade, surpassed all building records of the eighteenth century. An increase in the number and tonnage of vessels, the creation of new designs, and the use of new building materials were evident at the beginning of the nineteenth century. Copper, previously limited to the domestic manufacture of stills, kettles, and vats, became vital to ship construction.

When the wartime value of copper sheathing and other technological improvements was recognized by the British Navy, copper was elevated to the status of a vital naval store equal to that of munitions. To protect and maintain their own supply of copper the English first restricted its export in 1798. Other restrictions followed. The Americans, who depended exclusively upon the English rolling mills for sheathing, were the first to be deprived. The European war that increased the American shipping potential also brought about the English restrictions that transformed copper from a readily available commodity to one that was difficult to procure. This immediately affected American procurement, the carrying trade, and the rise of the nonferrous metal industry. The favorable conditions that characterized the importation of copper during the early 1790's had been brought to an end when Harmon Hendricks resumed his father's former business.

Up to the present time information relating to the supply of copper and the beginnings of its manufacture in the United States has been meager. The business records of Harmon Hendricks make possible for the first time a more detailed examination of an obscure branch of the nonferrous metal trade. The copper trade was so intertwined with the complexities of Anglo-American relations prior to the War of 1812 that a study of Hendricks as a shipping merchant specializing in copper involves an examination of the conditions that led to the war and the subsequent rise of the American copper industry.

Prior to 1800 the American demand for ship sheathing was limited to the coppering of a few ships. Coppered ships sailing into American ports became the envy of the shipping merchants and set an example for them to follow. As knowledge of the effectiveness of sheathing spread, the use of copper for American ships increased. The alert shipping merchant, eager to protect his vessel, adopted the new method at once. To obtain the necessary copper for the shipwright and the coppersmith Americans continued to import it from England.

Other circumstances that compelled the Americans to turn to the English mills were caused by the lack of domestic ore. Copper mining in the United States was still a minor industry, and where it was pursued—in Connecticut, in New Jersey, and in Maryland —its production was insignificant. A limited knowledge of metallurgy restricted attempts at copper refining and kept those industries dependent upon nonferrous metals from expanding. The craftsmen and the metalsmiths were also dependent upon the English manufacturer. This was to the advantage of the British smelters and rolling mills, who enjoyed the ready market in the former colonies for pig lead, copper, tin, and spelter, and for the articles manufactured from these metals, which could not be produced in the United States. By guarding her knowledge of technology Britain was in a unique position to protect her manufacture of brass and copper utensils, the drawing of iron wire, the rolling of sheet and plate copper, and the preparation of spelter or commercial zinc.

Between 1784 and 1798, because of the rapid growth of English industry, Americans were given a greater opportunity to purchase a wide range of goods of such excellent quality and price that they were dissuaded from launching a program to develop their own manufactures. No thought was given to the production of copper when it could be freely imported from the Bristol mills in pigs or sheets. Looking to the commerce across the seas, the Americans turned their energies to the import-export trade, offering agricultural products or choice securities in exchange for essential metals.

However, dependence upon English manufacture was precisely what public-spirited Americans feared. Although they were more concerned with establishing the production of cotton and woolen fabrics, the dressing of skins and leathers, the manufacture of paper, and the preparation of flax and hemp than with the production of nonferrous metals, they recognized the obstacles confronting the metal trade. Americans became major consumers of English manufactures, and the shipping trade that encouraged foreign goods to pour into the United States weakened the pleas for domestic manufacture.

Fully aware of English technological superiority foresighted Americans attempted to change this situation. Among the few in the metal trade who stepped forward to apply their mechanical genius and business skills to industry, and who had the courage to invest in manufacture, were Nicholas J. Roosevelt, Paul Revere, and Harmon Hendricks. During the first decade of the nineteenth century their efforts helped release the forces of technology and invention dependent upon nonferrous metals.

Roosevelt had determined to rehabilitate the Schuyler copper mine, once the most productive mine in the American colonies. It was located somewhat north of Newark, New Jersey, and before the American Revolution it had been successfully exploited by the British. Roosevelt applied his knowledge of engineering to mining and milling, convinced that ample copper could be mined to satisfy American needs. In his effort to find financial support to develop this enterprise he was confronted by England's attempt to gain control of the world copper market.

The rich Anglesea (Anglesey) mine in Wales had at this time exceeded all others in copper production, and its proprietors, in an effort to "monopolize the copper trade of the civilized world," attempted to annihilate all rivalry by underselling every competitor and by purchasing abandoned mines. But the Anglesea treasure was suddenly exhausted, and its decline provided the British government with another reason to prohibit the exportation of copper. Application was even made to contract for all the copper that the small American Schuyler mine could produce. Roosevelt was anxious to build a mill that could be fed by American copper, and he rejected the English application. Despite all of his efforts the mill did not meet with the encouragement necessary for its success.

In Massachusetts Paul Revere, the old patriot of the Revolution, was faced with a lack of encouragement when he founded his rolling mill in Canton in the fall of 1801. Revere was unable to obtain sufficient block copper and he had to rely upon brittle scrap and old copper. Frequently he augmented his supply by importing or purchasing it from Harmon Hendricks. Revere's interest in rolling copper was intensified when the Navy of the United States became aware of its importance.

Between 1801 and 1803 repeated demands were made for copper sheathing by the vessels that touched the harbors of Boston and New York. Revere supplied copper for the now famous *Constitution,* and Hendricks made available sheathing for the *Elinor,* the *Mexicana,* and the *Aspasia,* ships that required as much as 9,000 pounds of copper at a cost of more than $7,000. At about the same time, Navy purchasing agents, who Revere had hoped would support his mill, contracted for more than $12,000 of English block copper and bolts from Hendricks. If Revere feared a loss of naval work to his New York competitor there was as yet no indication of it. For Hendricks it was the beginning of a series of naval contracts that were to continue throughout his lifetime.

Revere's correspondence with Hendricks, concerned primarily with the scarcity of copper and other nonferrous metals, indicates some of the procurement problems faced by one of the first copper rolling mills in the country. His unfamiliarity with the English market compelled him to turn to Hendricks for pig lead and soft-rolled British sheathing, which Hendricks imported from Bristol. In 1803 both men were disturbed by the irregular supply, and Revere sought copper wherever it could be purchased in order to keep his small mill in operation. The practicality of combining light with other sheathing for vessels—at Hendricks' recommendation—was questioned by Revere, who had not previously used this method. Perhaps for this reason, or perhaps because of his curiosity about the New York trade, Revere decided that a "personal conference" with Hendricks was necessary. Trips between Boston and New York supplemented the two men's correspondence, as Revere attempted to meet the problems of copper manufacture in the United States.

The American demand for nonferrous metals convinced Hendricks that the lack of an adequate supply was a serious problem for both the shipbuilder and the craftsman. Revere, as Hendricks was soon to realize, could not possibly manufacture enough copper without dependable sources of raw material. England was the major producer of manufactured copper and controlled many of the sources of raw ore. But once England was confronted with the wartime crisis and had to keep her Navy seaworthy and her home industries constantly supplied, she legislated to classify the use of copper, to control its distribution, and to limit its export. Long before copper was subject to the regulations of the American tariff, English export restrictions curbed its supply.

The history of the restrictions affecting the export of copper to the United States begins in 1799 when Thomas Holmes, the British commission merchant who had been employed by Uriah Hendricks, reported to Harmon his first encounter with the laws affecting copper exports. These restrictions, which also included other items, were to continue in varying degrees until 1805, when the war across the seas entered another phase that affected the whole carrying trade.

The Orders in Council, as the restrictive measures were known, compelled English factors, shipping merchants, and manufacturers to obtain a costly export license, which could be used only when the copper ban was lifted. Following Holmes's report, further news of English export restrictions reached Hendricks. Stratton and Gibson informed Hendricks that "The Merchants intend petitioning to be allowed to export" copper. John Freeman and Company, whose smelting house and rolling mill had achieved a position of prominence in the world copper market, gave a different account of the same problem, explaining that copper could be supplied occasionally but only in moderate amounts. The English home trade, they added, was now paying full price, but regardless of the cost of copper a government export license was necessary.

It became clear that the Orders in Council, when first enacted, were not intended to deprive the Americans, for the ban was lifted in six months. Then the copper ordered by Harmon Hendricks set sail under the escort of an English convoy to protect it from seizure by enemy ships. Exports were therefore possible, but irregular and unpredictable. Hendricks was unwilling to rely upon chance and kept sending advance orders to the English copper firms in the belief that some of the orders would be filled before the bans were restored. And they were. Hendricks rejected offers to import goods illegally by evading the restrictions. For the next three years Parliament, still governed by naval considerations, carefully guarded all copper exports. In 1802 Parliament imposed further prohibitive measures that related to the refining of copper shipped to England from America. Freeman advised Hendricks:

The duties appear payable by Weight, the only Risk therefore is of siezure if any of the Copper prooves new or not to require being remanufactur'd, or if the Blocks appear to be of the Spanish colonies or of any Country except the American States to which the Ship belongs . . .

Later that year Freeman reported that Parliament was studying the consolidation of all duties on copper export and the enactment of new regulations. Gradually it became clear to the merchants on the American side of the Atlantic that their colleagues in London, Bristol, and Liverpool resented Parliament's interference with trade, a resentment which was to grow from year to year.

To this irritating circumstance was added England's loss of the European market, a result of the war with Napoleon. The market thus left open became a field for American commerce to exploit. A zest for commerce sent American-built ships across the ocean lanes of the world and, while England was busy vying with the French, the Dutch, and the Spanish, the Americans penetrated the European market. To fully sustain this lucrative commerce, American capital was poured into the shipbuilding industry and the carrying trade.

When the war with France entered a new phase in 1803 England could no longer restrain her concern over the remarkable progress of American maritime commerce. His Majesty's Navy was determined to sweep American vessels from the seas. With its overwhelming superiority the British Navy tightened the blockade of European ports, encircled the islands of the West Indies, increased the impressment of American seamen, and cut the supply lines that reached her enemies' ports. Depriving the French and Spanish of the staples carried by neutral American shipping was a necessary step in Britain's war against the belligerent powers. Only by restricting the neutral shipping rights of the Americans was this possible. Between 1804 and 1806 England issued a series of sharp restrictions calculated to destroy the rich American trade with the West Indies and other European colonies. The "Rule of 1756," forbidding neutral trade with the colonies of belligerents in time of war, was now fully invoked.

In order for the Americans to overcome the problem of shipping goods from the islands or from South America directly to Europe, and to assure the safety of the goods, they were sent to the United States and then reshipped, breaking the direct voyage. Involved was the principle of the "broken voyage," upheld by the British since 1793. With shipping orders issued from an American port, the same goods were reexported to European countries. Every effort was made to show that the goods were intended for American consumption. In this manner, direct trade between the colonies in the West and the European powers was avoided, and American neutrality maintained. When vessels engaged in this trade were seized and ordered before an admiralty court, the master had to prove that his voyage was not direct or, failing to do so, have his cargo condemned.

British toleration of the "broken voyage" resulted in bringing almost all of the carrying trade with Europe into the hands of the Americans. With the exception of Peruvian copper assigned to American ports, the "broken voyage" conveyed other goods from South and Central America and the island colonies to European ports. In return, the products of the Dutch looms that Hendricks bought from Daniel Crommelin, or the spelter that came from Germany crowded American shelves and supplied American metal craftsmen. South American copper was in too great demand to be reexported to Europe and was sent there only to be smelted and refined.

The principle of the "broken voyage" proved so harmful to the British that in 1805, following the seizure of the American *Essex*, it was reversed. Now the masters of American ships had to be supplied with indisputable documents that their voyages would terminate in an American port and no goods would be reshipped. Failure to do so meant seizure of the cargo by the British. In September, 1806, news of this decision reached New York, and at once the shipping merchants envisioned the doom of their rich trade.

This sweeping move was the first step on the road to war between the United States and Great Britain. The manner in which it affected the American procurement of copper, the importance of which has been completely overshadowed by the maze of political maneuvers between the two countries, has not been previously recognized. Ever since copper sheathing had been classified as a material of war it had become more difficult for the Americans to obtain. To curb but not entirely eliminate its export from the English refineries and rolling mills, the British once again issued new export restrictions.

At times when the English summers became unseasonably hot and water was scarce, copper mills were forced to halt, all of which further reduced copper production. Yet despite the complexities wrought by war, legislation, and nature on both sides of the ocean, large quantities of copper crossed the Atlantic. However, it was an irregular supply because shipment was difficult to arrange, prices fluctuated, and the rates of maritime insurance were governed by the misfortunes of war; and so competition for the metal was unabated.

Recognizing these erratic circumstances Harmon Hendricks showed his capacity to scrutinize all the English sources of copper and obtain a supply. His success could be seen in the tremendous upsurge in copper purchasing during the four-year period from 1799, when the first Orders in Council were introduced, to 1803 when the restrictions were sharpened to further reduce exports. He spent more than $100,000 with John Freeman and Company for old, raw, and prefabricated copper. In addition to these purchases from Freeman of Bristol, the bottoms and sheathing, the bolts, bars, spikes, and nails shipped by the manufacturers from Hull, Liverpool, and London showed the care with which Hendricks tapped the English sources. Hendricks was becoming the acknowledged importer of copper in the United States, and although he had assured himself a constant supply of metal it was difficult to meet the naval demands and satisfy the needs of the coppersmith.

Turning to his London agent, Thomas Holmes, he asked, "What can you do to induce me to order £5000 stg. Spring, £5000 stg. fall in copper?" With an insistence that revealed his determination to keep a constant flow of copper to New York he proposed to the Freeman firm that if they would undertake to supply his copper needs he would assign all of his orders to them. At the turn of the century Freeman rejected the first such proposal. When Hendricks repeated it in 1803 it was again rejected. A year later, Hendricks, unperturbed by two refusals, made the same request, and the owners of the Bristol mill politely rejected it a third time in the finely polished English prose of William Del-pratt, the head of Freeman's.

To further assure himself that he had reached all possible sources for the purchase of copper Hendricks gradually established a network of agents in the port cities along the Atlantic coast. Incoming ships with cargoes of unconsigned metal were bought on Hendricks' account. Old or scrap copper that could be found in any of the seaport towns was also an object of interest. However, such purchases required special attention, and in 1804 Hendricks purchased a foundry that could reconvert old and scrap copper. He informed his brother-in-law Jacob De Leon: "I have a manufactory established that can consume 200 tons of Old Copper a year" in the production of bolts, spikes, and shipwork. He gave De Leon a blanket order for old copper, scrap copper from ships, and new or old brass cannon, and advised De Leon that cash was available for any purchase not exceeding 22¢ a pound. Many purchases from many sources were required to keep the foundry on Bedlow Street working, but unfortunately the extent of Hendricks' first entry into manufacturing and refining copper is not known.

Although Jacob De Leon, who did purchasing on behalf of his brother-in-law, never became a full-fledged agent, Hendricks found it practical to engage others to join him in that capacity. David Moses of the New York mercantile family was among the first to do so. He became an agent in Boston. Others followed in Providence and, for a short time, Joshua Isaacs, Harmon's father-in-law, surveyed Newport as a potential source of discarded scrap copper from old ships.

During 1805, when the supply of copper continued to be irregular, Revere and Hendricks entered into an agreement for the joint purchase of block copper. The purchase was then divided and a price established. Thereafter the two men agreed not to compete with each other for copper consignments at the ports of their own cities. Unaware of this gentlemen's agreement, David Moses, acting on behalf of Hendricks, prepared to bid for a large shipment of quality pig copper. Hendricks advised Moses of the agreement with Revere:

My agreement with Mr. Revere was not to interfere but to let him exclusively have the opportunity to make his terms. And even if offered in our market I was to become purchaser on his own account. If this parcel should not be the one in question, or Mr. Revere has abandoned his first intentions, then I wish you to purchase the same on my acct. on the most advantageous terms & price not however higher in price than 22 Cents a pd, 21 cents for a draft on me at 60 days would do. I give it a preference.

Revere bypassed the opportunity to purchase the copper and Hendricks acquired it for some $6,000, maintaining a policy of buying every good shipment of copper offered in the American market.

II

Although copper was the core of his business Hendricks was deeply involved in many other branches of the carrying trade, and the principles that he applied to purchasing copper he applied to other imports.

Specialized items, such as gold leaf, were also a part of the Atlantic metal trade. Sheer chance brought Hendricks into contact with the London goldbeater John Atkins. Atkins, a young beginner, read Hendricks' advertisement in a New York newspaper that had come his way, and was induced to offer Hendricks books of gold leaf. For many years thereafter the neatly hammered Atkins gold leaf supplemented similar importations from Harford, Partridge & Company of Bristol.

In spite of the Orders in Council that controlled the export of a large variety of items, fine leather products were obtained in England where they were best tanned. The morocco that dressed books or lined clothing, or the kidskins for ladies' shoes, in "yellow, sky blue and pink," presaging modern colors, were made to order for Hendricks in England. An insatiable demand by the American country trade for white sheeting, toweling, diaper cloths, and other Russia goods kept that market lively. Pieschell and Brogden, known in the United States for their extensive connections in the Russia trade, represented Hendricks in the purchase of these goods. Between the spring of 1799 and the fall of 1802 Hendricks imported more than $55,000 in Russia goods alone. The safe transport of even these goods was endangered by the European powers' competing for the commerce across the seas. During 1801 the threat of a rupture between England and Russia over the Baltic trade brought about an embargo in the northern seas. Russia detained English ships in her ports, and England retaliated by placing an embargo on Russian, Swedish, and Danish ships. The Oppenheim brothers, who relayed this news to Harmon, had cause for concern, for the £2420 worth of merchandise they had shipped to New York in a ten-month period in 1801 had come largely from Baltic ports.

The wide range of goods that Hendricks bought in England —Italian and English taffeta umbrellas, choice cutlery from the London Da Costas, silvered mirrors from Pieschel and Brogden, and ladies' "Looking-glasses" from the Oppenheims—showed the extent of his New York market. Nor did he overlook the purchase of the popular steel and copper engravings from the Shakespeare Gallery, for which there was a constant demand in New York City. For his own use Harmon bought an excellent pianoforte from the Oppenheims, noting that a "cover of leather is usually given in the bargain." The piano must have been quite satisfactory, for soon after Hendricks placed an order for three more.

Much of Hendricks' success in obtaining English goods was dependent upon his ability to supply in return the staple products of the southern plantation—tobacco, cotton, corn, and molasses— all of which were in great demand by the English. Ships bringing copper to New York were eager to find a return cargo that could be easily sold, and the captains had little difficulty finding such goods with Hendricks. Hendricks had established good business connections with the brokers, commission merchants, and planters, who were as numerous in the South as his range of metal customers in the North. The southerners who formed this business unit were relatives of Hendricks and former residents of New York City. Through them Hendricks was put in touch with the sources for turpentine and pot- and pearl ash. He utilized all these business contacts to great advantage in directing a constant flow of goods northward in the coastwise trade.

Hogsheads of tobacco from Robert Pollock and Manuel Judah of Richmond were shipped to Hendricks in New York and exported to the Oppenheims in London. Bales of preferred upland cotton bought from Cohen and Moses and from Jacob De Leon of Charleston went north coastwise and were then transferred to a London copper agent, who forwarded them to the factories in Leeds. In 1804, to eliminate reshipment from New York, Hendricks had De Leon ship the cotton directly to England from South Carolina: 220 bales to Bristol, 250 bales to London, and 180 bales to Liverpool. That same year Hendricks' non-copper imports reached a high of $100,000.

Barrels of flour from Norfolk and Petersburg, Virginia, were also consigned to the North, and puncheons of arrowroot from Charleston were added to the cargo for a prospective Hendricks customer. The Charlestonians, although they had a port of their own, were eager to do business with Hendricks. Prior to the War of 1812 a large part of the goods exported by Hendricks to England came from the Carolinas and Virginia. The New Yorker remained cold to his brother-in-law De Leon's description of $70,000 in "Black Birds," slaves brought in from the west coast of Africa, and he was unimpressed by Sam Grove's boastful account of a cargo of 700 Africans who had been "packaged" in three vessels. Although he handled the bills of exchange, Hendricks had no taste for the slave trade and thereafter discontinued handling similar bills.

Wherever possible Hendricks avoided the importation of articles whose manufacture had been begun in the United States. To Daniel Crommelin of Amsterdam he reported that the sale of Dutch linens in the United States was without profit; and Walker, Bulmer and Horner of Leeds were informed that the woolens that Uriah Hendricks had sold profitably in the 1790's were now, in the first years of the nineteenth century, of small benefit. Competition from the looms and spinning mills of New England had reduced the importation of these products. Nonetheless, most of the goods required and desired by the Americans were not available in sufficient quantity in their own country to affect the carrying trade.

III

The distribution of the goods that reached New York was as painstaking as their procurement was difficult. Local tradesmen, the foundrymen, and craftsmen obtained their goods without difficulty. Bulk goods were delivered by carters who hauled them from the Hendricks' warehouse to the purchasers' shop front, where the goods were deposited.

There was no difficulty in reaching the city's twelve brass founders, fifteen coppersmiths, and twenty tinners, all of whom appear in the Hendricks' ledgers, or such specialists as the millwright Robert McQueen and the furnace operator John Youle, who regularly bought their metals from Hendricks. In addition to the craftsmen who comprised the local metal trade, the shipwrights were important customers. Of the shipwrights who did the coppering in New York, Foreman Cheeseman, believed to have opened his yard in 1792, was one of the outstanding shipbuilders who was an early purchaser of Hendricks copper. He was followed by Charles Brownne, who had mastered the shipbuilder's art in the British dockyards. In 1805 two new yards, built by Adam and Noah Brown, and Eckford and Beebe, were bustling with activity at Corlears Hook on the East River. In these yards the character of shipbuilding and naval architecture was to undergo great changes in the first quarter of the nineteenth century.

The inland shopkeeper, usually located on one of the water arteries that led to New York, was not so easily supplied. Nor were the merchants along the coastline who depended upon fast-moving packets that sailed the Atlantic waters. Early in May, when the country trade came to life, the sloshy, muddy roads had hardened enough for the teamsters to drive their wagons to the inland through a steady spray of dust. The advent of spring also opened the rivers of New York and Connecticut to navigation, and made possible the retail and wholesale distribution of Hendricks' extensive imports.

In the Albany hinterland Myndert Lancing, a Hendricks customer since the 1790's, awaited his copper, pewter, tin-plate workers, and skilled coppersmiths. Men, looking for steady country employment at city wages, were offered free transportation if they decided to remain at one post for at least three months. Hendricks found skilled labor for his rural customers, and supported the laborers' families who remained in the city. Repayment for this service was expected only after the mechanic was settled in his new job or if the family moved to the country. In this manner the shipping merchant became directly involved in spreading crafts and light industry to the inland; with the sale of metal goods skilled labor had to be supplied. Overcoming the scarcity of labor was essential to the distribution of metals.

Lancing was no exception. In Hartford, James Ward, a Connecticut Yankee, maintained a similar business relationship with Hendricks. Ward was a recognized silversmith, a metal fabricator, and an agent for Hendricks, but he was also a merchant in his own right. As such he was one of Hendricks' most active customers in the purchase of copper, tin, spelter, brass, and lead. Ward also called upon Hendricks for trained mechanics to come and turn the sheets of copper into kettles and stills, and to shape tin into plates. Ward knew the farmers in the area who worked for him as tinsmiths in wintertime and the city folk who sought his employ in the spring equally well. He lived in an economy of country pay and credit, and such a relationship better enabled him to distribute his goods throughout a wide Connecticut area.

Products made of tin were especially in demand by every American household. Tin and tinware became the backbone of the Yankee peddler's trade. Hendricks' entry into this market was made possible by the information supplied to him by Ward. This best explains his importation from Harford, Partridge & Company of more than $100,000 of tin-plate in a three-year period. At the beginning of the nineteenth century, when the tin peddler's cart was a familiar sight from the St. Lawrence River to the Potomac,

Hendricks supplied the tin and tinware carried in the carts of James Ward's corps of Connecticut peddlers, or those of Seth De Wolf.

Ward's Yankee ingenuity and practical thriftiness commanded the attention of Harmon Hendricks. From Ward came some of the working knowledge of how to treat damaged tin, which might otherwise be abandoned or sold for scrap. The damp holds of vessels were destructive of metals susceptible to rust. Rough seas and frozen harbors made tin a special target. Ward had his own way of salvaging a cargo of tin that had been submerged in water when a vessel was driven against the rocks by ice. Fearing that the fast-spreading rust would ruin the tin before it could be returned to New York for credit, Ward employed:

twelve labourers in rubing carefully with rotten stone, after rinseing and drying it rubed it in flour to make it perfectly dry, I do not know but that the small appearance of its having been rubed will injure the sale of it, but believe it is in reality as good as before at least it will not grow worse, was four & half days doing it . . .

Hendricks was charged with the cost of salvaging the tin, and he gladly absorbed the expense rather than have the tin returned.

After 1798 commercial transactions between Ward and Hendricks increased yearly and, as they did, the correspondence between the two men became more intimate. The aggressive but frugal Ward held the same moral views of society as his New York friend. Both men eyed gamblers with disdain and looked upon drunkards as outcasts. In the gloomy year of 1802 when they exchanged many of their views, an uncommonly high number of business failures took place, affecting both Ward and Hendricks. But each man was liberal toward creditors who were personally respectable and who had defaulted as a result of circumstances beyond their control. Trips to New York or Hartford were also social occasions for the two men, and on occasion a bit of pleasant family gossip would be added to their otherwise strictly serious correspondence about metals. Hendricks sent word of the birth of a son, and news from Hartford told of Ward's wedding.

Ward kept Hendricks informed of his new accounts in the Connecticut area and advised him of those he thought were financially responsible. A reference from Ward often helped a shopkeeper to establish credit. These confidences induced Hendricks to write Ward about a mutual customer who had defaulted in the payment of a bill. When the debtor decided to make payment he deliberately chose the Jewish Sabbath to do so, knowing that Hendricks would not conduct business on that day. Fully aware of the ruse, Hendricks was greatly annoyed.

The same spirit of friendliness and cooperation is found in the correspondence between Hendricks and Lancing of Albany. There is a noticeable absence of political comment between Ward and Hendricks, but Lancing expressed his concern about Aaron Burr and Alexander Hamilton, suggesting his preference for Burr. His Federalist views were no secret. However, Lancing, like Ward, was more often concerned with the personal care and attention that Hendricks gave to the immediate problems of shipping, the weather, the speed of delivery, and the quality of the goods that he ordered. At one time, in writing about the bitter winter of 1803, Lancing described the grain stores as "a tumbling down with the heavy loads [of] wheat [and] hay" under the strain of four to five feet of snow. That coming spring he looked forward to receiving his goods with unusual eagerness. The confidence and cooperation that Hendricks established among the inland merchants grew from year to year and was broken only by death. Such associations were vital to his success and were an important factor in the distribution of his imports.

Other metal tradesmen who relied on Hendricks for their supply of spring goods were found in the principal towns of New England, the middle Atlantic states, and the South. Seth De Wolf of New Haven and Thomas K. Jones of Boston received their goods coastwise. Jones had been a distributor of copper in the Boston area before Revere established his rolling mill in 1801, and was one of Hendricks' first accounts in Massachusetts. For decades Jones enjoyed the same warm relationship with Hendricks as did Ward and Lancing.

The importance of the New England trade made it compulsory for Hendricks to call frequently upon his customers in Boston, Newport, Providence, and Hartford. Mills were springing up in

The ledger account of Harmon Hendricks with Paul Revere.

these cities and in nearby towns. An interest in metals that had not existed before was soon apparent. After visiting Boston for the first time Hendricks rode to Canton, on the Neponset River, to see the "miniature manufactory" established by Paul Revere for the rolling of sheet copper. He quickly recognized that Revere's water supply for the mill was inadequate for a successful all-year operation. He recalled the heat of the English summers which consumed the English mills' water reserves, temporarily halted their operation, and cut the production of copper. During the first troublesome years of Revere's rolling mill he appeared to be considerably dependent upon Hendricks for copper, block tin, spelter, and pig lead. New England shipyards made demands for copper that Revere could not meet, notwithstanding the importations that supplemented the products of his mill. Hendricks therefore carefully allocated his Boston sales, bearing in mind that Boston was still many weeks' shipping distance from New York.

Many of the English copper bottoms that Hendricks sold to Revere were originally intended for stills, but instead Revere melted them down and rolled them into sheathing. This simplified his problem of refining mixed ore, or the partly refined blocks. The reconversion of these bottoms resulted in the production of sheathing whose quality Revere claimed to be equal to that rolled in the English mills.

There was a considerable interchange of sheathing copper and still bottoms between New York and Boston. Revere offered Hendricks quantities of spikes and nails of his own manufacture, and at one time ten tons of Mediterranean copper for which he had no immediate use. Hendricks in turn offered Revere the partly manufactured English bolts from which spikes and nails were cut. Business between the two men lagged only when each was plentifully supplied with copper, or when the yellow fever disrupted the normal commerce of New York. Then young Solomon Isaacs, Hendricks' apprentice, would note in writing a copper transaction, "you may find an asylum from the Flies, but not from the mosquetoes," unaware of the connection between the mosquito and the fever.

Hendricks and Revere competed for the same market, but Hendricks' customers preferred to buy their metals from him when they bought their other merchandise. The sale of other goods gave Hendricks a greater advantage in distributing copper to a larger number of customers. Occasionally competition for government work from another source became a matter of mutual concern. This was disclosed in Hendricks' report of a naval contract that slipped through the hands of both copper men. Government contracts for copper were still few and the copper merchants kept alert for every sale. Revere suspected that the successful bidder was Nicholas J. Roosevelt of the Soho Company, and when his suspicion was confirmed he felt concern for the future of his own rolling mill. Roosevelt made available enough bolts and spikes for a seventy-four-gun ship, all of which was made from American copper mined in New Jersey.

Revere was also concerned with competition in the New England market, because he feared the loss of manufacturing contracts and a regular supply of quality copper. Although Hendricks had assured Revere that "No rivalry can be expected between your & our markets," Revere did not hesitate to encroach upon the New York market, despite an agreement with Hendricks not to do so. According to James Ward, the old patriot visited Hartford with the object of selling the copper rolled at Canton:

Revere talked largely about the copper he was manufacturing for New York. We wanted to take 2 or 3 of his Sheathing to see if it would work as well as Imported but he declined & said he would send us some for trial that was rolled both ways in the mill—this being rolled only one way.

Hendricks' extended business trips to New England were adjusted to meet the requirements of the Jewish holidays, when he would not travel. Frequently a delay caused by impassable roads or rough sailing conditions would bring him to Newport for the Jewish New Year. He was among the last to worship in the city's historic synagogue when the circumstances of a declining Jewish population compelled it temporarily to close its doors. From Newport he would ride to Providence, spend a day with the trade, and then resume his journey. On one occasion he made the trip from Providence to New York riding horseback day and night through a dismal shower of rain.

Trips to Philadelphia were less frequent, although Philadelphia, as a rising center of American industry, required special attention to meet the growing competition in the metal trade. The Philadelphia shipping merchants were great consumers of copper because they were heavily involved in the general carrying trade, and engaged in an immense trade with the West Indies. The industrial metropolis was a major seaport and had its own navy yard, with such a potential that it required constant attention. When Hendricks decided to continue in his father's business, the gossipy but hard-working John McCauley became his Philadelphia agent. Benefiting from his past experience in distributing cigars, coffee, and island produce through various Philadelphia agents in the 1790's, Hendricks now applied the same sales principles to metals when he employed McCauley. McCauley, because of his knowledge of metals, took great pride in the trade, but he was not as aggressive as the firm of Beck and Harvey, the leading copper merchants of the city nor was he as shrewd as Vanuxem and Clark, a firm that sought the trade from western Pennsylvania to the Kentucky backwoods.

It was McCauley's task to advise Hendricks of the metal market in Philadelphia and act as a commission merchant for the retail and wholesale distribution of copper products. He had difficulty in meeting local competition, selling copper to the Mint of the United States, keeping abreast of Beck and Harvey, and parrying with competitors in copper price feuds. Hendricks was patient with McCauley's ineptness, but when McCauley was struck by reverses in his own business he relinquished his agency for Hendricks. When McCauley stepped down Solomon Moses replaced him. Moses, after his marriage to Rachel Gratz, had moved to Philadelphia in 1806 to represent the interests of the wealthy Moses family of New York. He combined his own business of commission merchant and auctioneer with that of representing Hendricks in the Philadelphia copper trade. For more than forty years Moses acted for Hendricks and his successors in the copper business, coming in contact with all of Philadelphia's metal merchants and manufacturers.

Solomon Moses was far more successful in Philadelphia than his brother David in Boston. He made excellent use of his valuable Gratz and Etting family connections, and through the Ettings he reached the Baltimore market, which was just beginning to compete with the markets of Philadelphia and New York. Moses' position as a commission merchant enabled him to remain independent while he negotiated Hendricks' accounts. Besides Moses, Philadelphia's thriving commerce attracted a number of other New Yorkers, one of whom was David Gershom Seixas, the son of Shearith Israel's minister.

Seixas actually began his hectic and unusual career as a copper agent for Hendricks. It was Isaacs, Hendricks' brother-in-law, who brought the two men together in business. Setting aside all formality Hendricks stated his terms to Seixas, described the rich potential of the Philadelphia market, and the rate of commission he paid to his agents. Seixas accepted the terms and entered the competitive Philadelphia market which, obviously, Solomon

Moses could not master alone. Seixas lost no time in establishing his new business by advertising brazier and sheathing copper in the Philadelphia press. Orders for bolt rods, manufactured by Revere in Boston, were shipped directly to Philadelphia where, for the next nine years, Seixas showed his capacity to participate in an expanding industrial area. During this time he became involved in the study of the deaf and dumb and, absorbed by this new interest, he eventually founded the Pennsylvania Institution for the Deaf and Dumb.

South of Baltimore, where Hendricks was engaged in a diversified type of trade, no agents were as yet required. Copper and tin-plate were not in the same demand as they were in the growing industrial North. A flourishing economy dependent on the plantation system and the products of the soil enabled Hendricks to acquire goods for reshipment to England, but his sales to Virginia and the Carolinas were limited to small orders for copper. The request Hendricks received from a Richmond firm for sheathing to be used for a drawbridge was not typical of the use of copper in the South. Sales of copper for stills to Newport or Petersburg, Virginia, were more common, but much less so than similar sales in cities of the same size in New York State or in New England. New Orleans was an exception but it, too, stood apart from the southern trade. Jacob Hart, a former New Yorker, became one of Hendricks' first New Orleans customers and was active in the purchase of Russia goods.

Hendricks' system of agents and his reliance on commission merchants to keep him informed of local markets and of the arrival of unconsigned copper shipments was a valuable asset to the distribution of his goods. Those who came in contact with Hendricks may have been attracted by the personality of the man and his willingness to give a newcomer a start in business. There is strong evidence that it was a part of Hendricks' Jewish training to help young relatives obtain a foothold. The evidence describing his agents and commission merchants who constituted his unofficial organization is sufficient to provide a picture of the activity of a group of young New York Jews who entered the copper trade under Hendricks' direction. The Moses brothers, David Gershom

Seixas, Haym M. Salomon, the son of the revolutionary patriot, and Solomon I. Isaacs formed a small network for the sale and distribution of Hendricks' copper imports.

Of this group Hendricks' brother-in-law and apprentice, Solomon I. Isaacs, is an outstanding example. Early in 1807 Isaacs completed the terms of his apprenticeship and entered the employ of Hendricks at the annual pay of $400. Isaacs worked very much the same as Harmon Hendricks did when he was a young beginner, his training differing only in that it was more specialized. He was taught the metal trade exclusively, with an emphasis on brazier copper and the foreign metal market. His duties consisted of scouting for bulk scrap and appraising South American or Cuban copper at public auction. This became increasingly important when the supply of refined copper from England was sharply reduced. Isaacs showed his skill in identifying various grades of copper, as a purchasing agent, and as a general tradesman. A year and a half after he entered Hendricks' employ, young Isaacs was elevated to the position of an independent commission merchant. While he busied himself exploring the field for new customers he acted as a representative for his brother-in-law.

One of the accounts that Hendricks turned over to Isaacs was that of Paul Revere & Son. Joseph Warren Revere arranged with Isaacs to negotiate the sale of Revere sheathing and bottoms and to purchase unconsigned copper that arrived in New York harbor. Gradually, all of the transactions between Revere and Hendricks were transferred to Isaacs. Hendricks also directed Isaacs in the field of brazier copper, in which the Reveres themselves were specializing. Isaacs occasionally advised the Reveres about the lively Philadelphia and Baltimore trade, the market for braziery, and the wants of his employer. It was the beginning of a career that later brought him into copper manufacturing.

Isaacs was the most successful of the five agents and commission merchants who were associated with Hendricks. The capable training and guidance he received from Hendricks enabled him later to contribute to the trade with which he was to be associated for many decades.

With this small group of men, which could hardly be described as an organized business force, Hendricks was able to carry out and extend his business. He did his own bookkeeping and wrote the duplicate and triplicate copies of an extensive foreign and domestic correspondence. His agents worked independently and were free to accept any line of commission business in addition to metals. The plan was by no means a loose one. It was typical of the methods of the time, it permitted a wider latitude for the distribution of goods, which was often dependent on the good will and cooperation of agents, and enabled many of the new-comers to the trade to learn the business.

IV

In contrast to the growth of the coastwise and inland trade American trade with the West Indies had waned. Slave revolts on the islands and the war between France and England that reduced the island trade also brought about a temporary decline in the Jewish communities. The Jewish tradesmen and merchants with whom Hendricks had done business were adversely affected, and they either set sail for the American mainland or else returned to England or Holland. The Gabays and Pesoas left for Philadelphia; Moses Da Costa sent his children to be educated in New York under Hendricks' supervision. Aaron Henriques hesitated with his plans long enough to witness the reestablishment of peace on the islands. Although there was a slight upsurge in trade immediately after peace was restored in 1802 the export of island produce which Hendricks once viewed with favor was now dependent upon the "broken voyage" and the risk associated with the indirect trade.

Several of Hendricks' island friends had meanwhile experienced a great deal of personal misfortune. The mother of the Da Costa children, who were living with Hendricks in New York, was burned to death when a wax taper ignited a vat of rum. David Pereira Mendez saw six of his seven children die at an early age. Only his seven-year-old namesake was still alive in 1802. Following the news of these tragic events the Gabays returned to Kingston, lamenting the loss of their slaves. Much of this news was contained in Mendez's letters to Hendricks, and in one of these

he introduced Jacob Levy, Jr., a pleasant man of "handsome capital" who had also left the islands for the promise of New York.
Levy and Hendricks were subsequently associated in many business ventures.

The flux of immigrants from the West Indies to America, the
first since the Revolution, brought with it the tastes and manners
of the islanders. Hendricks was not uncritical of the new immigrants. He sensed their discontent with the new environment
and their failure to adapt to the brisk manner of a growing nation,
but he was eager to help them. One newcomer whom he hoped
to establish in business was his London cousin Mordecai Gomez
Wagg. Wagg was said to "cut a dash," but before long he proved
that he had a greater flair for manners than an aptitude for business. After a few months in New York and the acquisition of more
debts than he could balance, he moved to Charleston to work
with the De Leons. Far more enterprising than Wagg was Samuel
Solomon, the Liverpool quack and medicine maker. His colorful
personality successfully aided him in selling his cure-all pills and
medical books to the residents of lower New York.

Most of the immigrants had neither friends nor relatives and
did not fare as well as Wagg, or meet with the success of Dr.
Solomon, or enjoy the financial resources of Jacob Levy, Jr. Moreover they disturbed the settled ways of the older residents, many
of whom no longer understood European mores, and this gave
the Americans cause for adverse reflection. "Europe hath emptied
her outcasts and the current of golden Expectations have drawn
them to America as numerous as shad in Summer, which very
often occasions an effluvia on the nation."

In turn, sharp criticism of the Americans came from the Eng-
lish-speaking Jamaicans whom Hendricks knew so well. The up-
rooted islanders were not happy about the loss of their possessions
and the difficulties that confronted them in earning a living. They
looked to America, but their thoughts remained with England.
Some of those who visited New York with the intention of remaining returned with an unfavorable impression, and those who settled there complained about American manners. To Aaron Henriques of Kingston, Harmon Hendricks expressed himself clearly
about the Jamaicans:

. . . their Turn of mind and uncultivated manners not preposing the better sort of our society having produced in them a disgust in our city & inhabitants that will no doubt be unfavorable in their reports, this situation you, my Dear Sir can never get into . . . my remarks are only leveled at a very small proportion of visitors from your place.

The children of the Kingston Da Costas were entrusted to the care of Hendricks in 1801, the year that his daughter Hetty was born. It was an occasion to rejoice, and the Hendrickses forwarded the news to De Leon in South Carolina, now the father of six children. A year later another son, Uriah, was born to the Hendrickses and they were delighted by the thought that the family name had entered another generation. Sol Levy of Peekskill also named one of his sons, born in 1802, Uriah, and De Leon was later to choose the same name for one of his children. Announcements of the births of their children were included in the exchange of business correspondence between Hendricks and De Leon, softening the formal tone of their letters. Information about bales of cotton and hogsheads of tobacco was secondary to the glowing account of Mordecai De Leon, who "acquitted himself with credit to himself and much satisfaction to me and his mother . . ." at his bar mitzvah.

There was a succession of sons born to Frances and Harmon within the next few years: Henry, Joshua, Washington, and Montague. In addition to Justina, whose birth preceded Montague's, seven daughters were born to the Hendrickses in the years that followed. As the family grew in number the Pearl Street home could no longer accommodate them. At 76 Broad Street a spacious brick house was rebuilt to Harmon's specifications, and the old residence was converted into business quarters. The Water Street offices and warehouse remained as before. Even the new quarters became crowded at times. The young brothers and sisters of Major Mordecai Manuel Noah were sent there to board. Harmon's sister Sally moved about as a strange and quiet observer, and the Isaacs family were regular visitors. Solomon I. Isaacs, the nineteen-year-old brother of Mrs. Hendricks, had just entered the copper trade in the employ of his brother-in-law, and was frequently present at the Broad Street house.

There was time to roll in the grass with the children and time for a ride through the country in the family coach with Harmon at the reins. Accounts of the children would follow Hendricks on his trips away from home. While Harmon was on a visit to Paul Revere in 1805 Frances devotedly forwarded details of the children's health, conveying little Uriah's questions about the mode of travel:

Hetty recovers fast as our fond hopes could wish, our little Henry is better, Uriah is thank god, in perfect health. I cant get him to understand your mode of traveling, he dose not know what a stage is, asks if you have your own horse. Solomon [Isaacs] sleeps here, dont be under any anxiety respecting the house. I take care that it is locked and the fires put out before I go to bed. I need not tell you how much I wish your return as I am sure you will not remain one hour after you have accomplished your business.

Harmon was able to relax with his family because he could depend on Solomon Isaacs to handle the mail, attend to banking matters, and recognize the difference between a bar of Peruvian copper and one that was English. A fall vacation in Newport was always looked forward to with great anticipation. It gave the family the opportunity to avoid the yellow fever that returned to New York with annual regularity. The Newport synagogue no longer mustered a quorum of worshipers, but there was a chance to visit some New England friends or one of the resident Lopezes. When the family returned to New York they lived in Greenwich Village until the last fall vessels arrived from England and the colder winds assured a healthy return to lower Manhattan.

There was time for other interests as well. Hendricks was among the first supporters of the Society for the Reformation of Delinquency, a problem about which he surely was not an optimist, for he became a lifetime member of the society. With a similar civic interest he joined the First Regiment, New York Militia, in 1804. When summoned before a court-martial for nonattendance at a company parade he refused to appear because it was on the Sabbath, adhering to the standards of strict religious observance. He contented himself by paying the customary fine for absence.

His association with Shearith Israel was as intimate as his father's had been before him, and his financial support of the synagogue kept pace with his increasing wealth. Details of his charity accounts are only fragmentary and are recorded in annual sums that ranged from $500 in 1804 to $1,500 in 1806, amounts that were exceedingly liberal for the time, and which increased yearly. Many of the appeals to the Hendricks' bounty were private. Simeon Levy, a Hebrew master who found little profit in teaching the holy tongue to children whose parents preferred French and Spanish, was one of many compelled to turn to Harmon for aid.

The impoverished Levys had one regular visitor who relaxed in their household, and who was as indifferent to poverty as she was unimpressed by wealth—Sally Hendricks. On entering the Levy house for the first time Harmon was moved by the sight he later described. Abject poverty surrounded him. There were broken, three-legged chairs; old, rickety furniture; and six "great girls, all Cinderellas setting on the hearth, in the midst of the group our Sally sat on a chair unconscious of any difference between such living or better, such is the poor habitation of Sim Levy." Hendricks provided for him unstintingly.

Sally preferred to be left alone. Her intellectual faculties had not improved with womanhood. Her different ways, for which there was no explanation, set her apart as a semi-tragic figure. In a less ordered family, or without the help of an interested member of her immediate society, she would have been a castaway. In a poorer household she might have been sent to the local almshouse, to spend a lonely and neglected life. But her brother's eagerness to fit her into normal society, as far as that was possible, granted her mobility within the family and a small circle of close friends. In an attempt to keep his sister occupied Harmon planned trips for her. But when Sally visited the Charleston De Leons she was content to be away from home for only a short period. Sally returned to New York to spend her days as a strange figure, wandering from household to household.

The broad view taken by Harmon Hendricks revealed a sensitivity which was unusual for his time to this human problem. Along with his massive orders of goods came the latest English pamphlets on mental disease, which Hendricks studied with the same diligence with which he absorbed the latest prices current.

That the information he derived from this literature was of little avail in solving his problem mattered little. It was the affection and interest that he showed his sister and the efforts he made to make her life a comfortable one that mattered more. Compassion, rather than embarrassment or fear of the unknown, motivated Harmon Hendricks.

On a winter's evening, after dinner, when the children were asleep, Hendricks would select one of his sharp-pointed quill pens and finish the day's correspondence. By bedtime he might have smoked twenty little cigars, and he took pride in writing De Leon, "I am an old Smoaker." He must have puffed intensely when writing about Naphtali Judah, whom he contemplated suing. Legal disputes arising from business claims were not uncommon and, like his friend Ward, Hendricks avoided the courts. Wherever possible, a settlement that involved Jews was made through a properly appointed body of three arbitrators according to Jewish law; this was similar in method to the Quaker practice of arbitration. Naphtali Judah's bankruptcy cost Hendricks more than $15,000 and no suit was entered into, nor was the matter privately arbitrated. Judah, after being discharged from bankruptcy, was in debt all over again. Hendricks petulantly entered the sum on his ledger as a loss. Many of Judah's creditors were unable to absorb such losses, and this galled Hendricks all the more. After prodigious exertion Hendricks did manage to recover £121 for his cousins, the London Oppenheims.

Hendricks attended to the legal and personal affairs of English relatives whenever the need to do so arose. Aunt Hannah Mordecai, for whom the elder Uriah had sought a groom in the 1750's, was now old and ailing. Her New York nephew was appealed to for aid, but before he could respond Hannah died. Rachel Wagg, his mother's sister, was widowed in 1803. Shortly after, her brother, Abraham Gomez of New York, died, and Harmon became the administrator of the Gomez estate for his English aunt. Other unsettled legal problems of the Waggs also devolved upon him, and each was executed with care. Money was advanced to Aunt Rachel; correspondence was forwarded to her elusive son, Mordecai Gomez Wagg, who had meanwhile encountered more trouble with his creditors. Hendricks made efforts to extricate his cousin from the "sharp set fellows who will not sign off," and finally he obtained a power of attorney from Wagg to help him settle his mismanaged finances.

All of the gentility, thoughtfulness, and care that Hendricks extended to his friends and relatives, he ironically did not apply to himself. His public restraint, like his dignity in private, must have been severely taxed, as the cryptic entry in his ledger tells, "To cost of pulling Samuel Hicks nose, verdict, $946."

V

In the spring of 1806 Congress, in an attempt to halt British infringement upon the neutral commerce of the United States, passed the first Non-Importation Act. The act prohibited the importation, among other articles, of tin and brass, two major items on the Hendricks sales chart. Metal merchants, harassed by the reversal of the "broken voyage" and the Orders in Council that regulated copper exports, were now faced with restrictions intended to hurt the English but which deprived the Americans of basic commodities.

When the most carefully managed purchase of copper had been approved for export between the restrictive bans, Hendricks was fortunate to secure freight space on ships even at the high rate of 50s. per ton. Then, as each of his invoices records, he had to obtain insurance from underwriters who had become reluctant to issue coverage when shipping conditions were erratic. Perplexed men like Hendricks, squeezed by the English resolutions, limited by the Non-Importation Act, forbidden to import tin and brass, dependent on the good will of a friendly skipper for freight space, had to persist in their efforts to bring in the metals so vital to the expansion of American industry.

But the most harassing news that reached the copper merchants came in the renewed debate over copper duties by Congress. The duty on copper was a source of continuous annoyance to Hendricks because he was convinced that the moderate protectionist features contained in the laws of 1790 and 1792 were misinterpreted by the local customs officer. Copper bolts and flat bottoms were classed as raw material and permitted to enter duty-free, but the customs officer of the port of New York confused them with plate and sheet copper which were classed differently, and levied a duty on Hendricks' importations. The bolts, from which nails and spikes were cut, were clearly duty-free and Hendricks argued constantly against the ambiguity and confusion that prevailed at the local customs office. Although he was charged with the duty he was permitted to post bonds for the release of his copper until the difference between raw and finished copper was clarified. Hendricks pointed to the fact that the practice of collecting duty on bolts and bottoms had been abandoned at another port of entry, which he did not name. He was convinced that Congress would eventually recognize the misinterpretation of the law and the confusion at the ports of entry and return his bonds.

In the summer of 1805, when it became known that Congress was prepared to levy a duty on the copper that had hitherto entered free, all of the copper merchants were alarmed. Hendricks was especially concerned. Beck and Harvey, the Philadelphians, entered into correspondence with Albert Gallatin, the Secretary of the Treasury, remonstrating against the proposed duties. The Philadelphia merchants then asked Hendricks for information that might be useful in support of their plea. Uncertain about the possibilities of convincing Gallatin, Beck and Harvey hoped that the trade would join them in a common petition to Congress urging that copper remain free. Furthermore, Beck and Harvey were eager to obtain Hendricks' signature for their petition, and were anxious to have him solicit the support of the New York copper men in their behalf. Once this was done, Beck and Harvey intended to petition Congress in person.

Hendricks was strongly opposed to the legislation, but his reaction to it differed from that of the other merchants. As a manufacturer, and as the largest importer of copper in the United States, he was determined to present his own petition. The prospect of lending the prestige of his name to a petition drawn up by a competitor was not appealing. Beck and Harvey attempted to monopolize the Philadelphia market, and they were always a step ahead of Hendricks' former agent, John McCauley. Only when Beck and Harvey sent their final draft to Hendricks, defining their process of fabricating imported copper, did he act. On February 25,

1806, Hendricks addressed a petition in behalf of the five New York copper importers who would be affected by the law. In order to bolster the importance of his petition Hendricks obtained the endorsement of the foremost shipbuilders in the area, Eckford and Beebe, Adam and Noah Brown, Charles Brownne and Christian Bergh. Meanwhile Beck and Harvey attempted to persuade Baltimore tradesmen to join them in a separate petition. Suit was filed in federal court in the District of Maryland by still another merchant whose experience in interpreting the old law had been similar to Hendricks'. During April, 1806, the bill laying a duty on copper was read twice in Congress.

From New England Paul Revere addressed his own petition to Josiah Adams on April 13, 1806, just a few months after the New Yorkers presented theirs, strongly recommending a tariff on sheet copper. Revere proposed that old copper, from which sheets were manufactured, remain free of duty. He claimed that the difference between sheet and plate copper was confusing. Plate copper, he said, was the same as sheet, for which no duty was collected. All of the petitioners agreed that the regulations of 1790 and 1792 had been misinterpreted and had created hardships for the importer and the manufacturer. The petitions of 1806 came too late for congressional action, and the petitioners failed to convince Congress.

Early in 1807 Paul Revere & Son led the petitioners in another address to the Congress. The petitions of 1807 dealt specifically with the problems of a young industry and the role of the shipping merchant in making copper available when American mines could not, and reflected the competition among the copper men. In addition the Reveres sought government support for their rolling mill, which was then the only one in the United States, claiming that "they can supply the whole quantity at present imported into the United States . . ."

To Gurdon S. Mumford, a respected merchant, elected to the Ninth Congress for New York City, Hendricks stated his opposition to the manufacturing claim made by Paul Revere & Son. Hendricks wrote:

I observe with pleasure you are one on the Committee of Commerce and Manufactories, and as Col. Revere's petition has come before your committee, and taken up I request permission to call to your mind a truth you are acquainted with. Merchants in all parts [of] the United States under an idea of Coppering their Ships cheaper in Europe sent them for that purpose, the practice I take it would become general if Congress should impose a duty on the article. This of Course would totally prevent the importation and no revenue would result, Besides the shipwright would loose the advantage of part of their Trade. It would be deduced from Col Revere's petition that he could supply the wants of the United States with domestic copper. I do not hesitate to say it is out of his power to supply Boston market solely. His minature manufactory & works have not one third of the year the advantage of Water, I have been at Canton where are established those works and I write from my own observations. Admitting Col Revere could manufacture sufficient for home consumption, there exists an evil not yet remedied, the want of Raw material. You will recollect the failure of Roosevelt, Mark & Co to work the Schuylers mine afterwards taken up by the Bank of New York under . . . incorporation and abondoned as a luising undertaking. Expereance of 20 odd years in the Copper trade warrants my ascertion, that there is not old copper in the United States sufficient for the manufacture of household castings nails, and other ship work. The deficiency of this article to my own knowledge has at times been made up with new copper even when at a price of 50 cents as was the case with a ship of Leroy & Son.

Hendricks then drafted another petition representing the New York importers, who were anxious for a remission of duties on copper in round bars. In January, 1808, the Committee on Commerce and Manufacture, to whom the petitions were referred, issued their report in a special document replying to the Reveres and the copper tradesmen of New York and Philadelphia. Recognizing the stimulus given to copper manufacture by the Reveres, and noting the New Englander's claim "that they can supply the whole quantity at present imported into the United States," the Committee on Commerce and Manufacture carefully weighed their recommendations. A duty on sheet copper would only hinder the manufactures they were anxious to encourage. Hendricks' declaration about the inability of the Reveres to produce enough sheet copper to supply the American market may have influenced the committee to reply to the Reveres: "That so much of the petition as prays for the imposition of seventeen and a half per centum ad valorem as duty on copper manufactured into sheets, is unreasonable and ought not to be granted." The committee also doubted the ability of the Reveres to roll a supply of copper ample for the American market. Old copper, which all of the petitioners wanted to remain free of duty, was considered raw material, but the duty on copper in round bars was not clarified. The issue concerning the temporary posting of bonds, a practice begun in 1803, as a substitute for the payment of duty, awaited a decision of the federal court of the District of Maryland. Thus the matter of the bonds rested uneasily until a later date, and the relief granted from copper duties proved beneficial to the shipping industry.

Prior to this time the facts relating to the first serious attempt to levy a duty on copper have been treated lightly because the law was not enacted. In addition, the failure to recognize Revere as the major advocate of a tariff that would protect the only rolling mill in the United States, to the disadvantage of the shipping industry and the coppersmith, has obscured the circumstances that brought about the petitions. A copper tariff at that crucial period in American industry might have further slowed the development of the ship industry by compelling American ships to turn to the English yards for sheathing. As a result the English mills that controlled copper manufacture might have also determined the fate of the largest American consumer of copper, the shipowners.

VI

The tariff episode that provoked the copper men into issuing a series of protests was of minor significance in the total foreign commerce of America. Even Hendricks could not be distracted by the tariff when faced by the problems of a constantly growing shipping trade. Hendricks had not lost his boyhood ambition of becoming a shipowner, and he had nurtured the idea of commanding one ever since he had been in the West Indian trade. Owning a vessel became imperative when the shipping supplied by the spring and fall vessels was found to be inadequate, and arrangements for freight space often depended on the faithfulness of a reliable agent. If a ship with valuable merchandise was grounded and thereby delayed for many months, the best agent was helpless. Such was the case of the *New York Packet* after leaving Bristol: "in going down the River got aground & has injured the keel, which has obliged her to return to repair the damage . . . the next Spring tides we expect she will go to sea again." Meanwhile Hendricks' cargo of metal, although undamaged, was delayed for six months.

Following this incident his Bristol agent, Thomas Holmes, lost an opportunity for freight in an available vessel to a competing merchant. Competition for freight space was an aspect of the shipping trade that harassed every importer. Freight agents for the American carrying trade were as alert for a share in the hold of a vessel as they were for salable goods. The ten English and American ships that regularly plied the seas between the ports of England and New York City carrying goods for Hendricks—notably the *New York Packet*, the *Oneida Chief*, the *Brig Fair Trader*, the ships *Otis*, *Enterprize*, *Hardware* and the *Bristol Trader*—averaged two trips a year. Though the number of merchantmen increased from year to year the carrying trade was in advance of what the ships could transport. It was no longer the expression of a young man's dream but a practical necessity when Hendricks employed the services of Robert M. Steel to engage a vessel that he could use in the carrying trade.

The *Ophelia*, "a strong burthensome vessel" of 283 tons, built at New Bedford, Massachusetts, was three years old when she docked in the North River on September 16, 1806, as the jointly owned vessel of Harmon Hendricks and the former sea captain, Robert M. Steel. For the next forty-eight days painters, riggers, sailmakers, carpenters, and coppersmiths were busy making her seaworthy. The hull of the *Ophelia* was scraped and recoppered with sheathing of Hendricks' selection. She was stocked with naval stores and a fresh water supply, and employed a crew of six seamen, a cook, a steward, a carpenter, and a cabin boy. The New York *Gazette* advertised her sailing for Bristol early in November, a number of passengers were accepted, and only the cargo remained to be loaded. Stevedores wrestled with the barrels of assorted turpentines, and carters hauled more than 33,000 barrel and hogshead staves on board. These made up the bulk of the cargo that cost more than $5,000.

Hendricks advised his English friends, from whom he solicited return freight, that the *Ophelia* "rates at Lloyds as a first class ship." Orders were placed with the three Bristol firms who supplied Hendricks with his metal goods: John Freeman for copper; Harford, Partridge for tin-plate—whose import was still restricted by the Non-Importation Act—and brass; and Philip George for iron wire. These orders, Hendricks stressed, were for the *Ophelia* and bore no relation to the purchases consigned to him on other vessels. He was especially interested in obtaining the Bristol goods because fewer vessels sailed from that port to New York. The highly desirable turpentine bought from Aaron Lazarus of North Carolina was consigned to Perry and Hays, to be resold at their discretion. One hundred and twenty barrels of pitch were sent on speculation. In return Perry and Hays were to send Hendricks as much cutlery, utensils, revolvers, and assorted hardware as possible. Captain Thadeus Waterman was given the customary instructions that merchants prepared and Hendricks, who was the author of these, revealed his extensive knowledge of how to pack metal freight into vessels. His instructions were precise and his specifications for storage were exact. The thousands of sheets of copper, flat bottoms, and the many tons of assorted copper were to serve as ballast "stowed between Decks in the Wings." "Tin plates is an article of all others liable to damage by water and damp, you will therefore take the least liable place to damage them in the ship. Iron wire is likewise an article liable to rust." Careful instructions were provided for the detection of rust at the time of loading. Hendricks also advised Captain Waterman to avoid those articles whose import was now forbidden by Congress, such as ten- and twenty-penny wrought-iron nails. At the end of the lengthy orders, Hendricks added a single personal request:

Have the goodness in one of your leisure strolls along Bristol streets to pass in a Toy shop and purchase for me one of the best Leather dolls with eyes, hands feet & toes. The cost I will cheerfully reimburse you in New York.

On November 8, 1806, Captain Thadeus Waterman locked away the leather pouch containing his orders and the ship's mail, and set sail for Bristol under favorable winds. After a passage of twenty-eight days and an encounter with a fierce gale that tore away the lifeboat on the stern, Captain Waterman was able to report the safe arrival of the *Ophelia*. At Bristol he had little difficulty in consigning the desirable Carolina turpentine to Perry and Hays. The Bristol merchants reported to Waterman that they had not received any communication from Hendricks about return freight, and what freight they had was already committed to the ship *Venus*. Besides, the English merchants had reacted so adversely to the Non-Importation Act that they were not very happy about consigning any goods to the Americans. When this news reached Hendricks across the Atlantic some time later, he addressed Captain Waterman by means of a fast-sailing ship, advising him of what had just taken place in Washington:

I have only time by this conveyance to inform you the President of the United States has in his second message to Congress informed them there was every appearance of a good understanding between our Country & G. Britain and recommended them to suspend the act prohibiting Sundry goods from England—to a further period. I am of opinion you will be safe to solicit and obtain every kind of goods, to be had on Freight in our Ship Ophelia. I shall write you again, interim subscribe for R.M. Steel & myself.

Hendricks' order for tin-plate and brass was thoughtfully placed in advance of the suspension of the Non-Importation Act. The action of Congress removed some of the gloom from the shipping trade, but freight, beyond Hendricks' copper orders, was still difficult to find. An offer from Harford, Partridge to accept from 100 to 200 tons of the best English bar iron at a freight rate of 20s. per ton, 30s. below tonnage rates, was not attractive, but by the time the *Ophelia*'s cargo was loaded the ballast consisted of 7,791 bars and 2,927 bolts of iron from Harford, Partridge. There were more than 1,600 boxes of tin-plate, hundreds of tightly crammed nests of iron pots and brass kettles of all sizes, forty barrels of Spanish Brown, 127 boxes of tobacco pipes, 134 bundles of iron wire, and in addition to a variety of ironmongery, castings, chains, and agricultural implements, there was more than $15,000 worth of copper from John Freeman and Company. Through the efforts of Captain Waterman and the exertions of Perry and Hays, the return cargo to New York was a rich one. Perry and Hays felt that their prior allegiance to the owners of the ships *Enterprize* and *Venus* would prevent them from assembling a similar cargo for the fall.

Captain Waterman was not without his own difficulties while he engaged the coppersmiths, the sailmakers, and the ship chandlers to put the *Ophelia* in condition for the return voyage. The boatswain was impressed on a British man-of-war, several of his seamen deserted, and the one-shilling-a-day cook earned himself a severe flogging for drunkenness. To complicate the problem of the return cargo, a small portion of it was destroyed by fire at the time of loading. On March 2, 1807, the *Ophelia* set sail under a British convoy to evade the French blockade and, after a voyage of seven weeks, arrived safely in New York.

With all the difficulties encountered by the *Ophelia*, a profit was made on her first voyage, and Hendricks was successful in getting metal goods that could not otherwise have received. No time was lost in outfitting the *Ophelia* for a spring voyage, but this time her destination was Amsterdam, not Bristol. The cargo consisted chiefly of large quantities of pearl ash, bags of coffee, bales of cotton, logwood, and thousands of barrels of sugar that were bought from Sam Grove, the Charleston slave trader who successfully defied the English blockade of the West Indies to obtain it. By reshipping it to New York, the "broken voyage" was legitimatized and the shipment was thereafter direct. In the middle of May the *Ophelia*, laden with a large cargo, set sail for Amsterdam. At Amsterdam the cargo was consigned to Messrs. Courdere and Brantz, who were to give Captain Waterman a letter of credit for £320. From Amsterdam the *Ophelia* was instructed to go to the port of St. Ubes, where she was to receive a full cargo of salt and return to New York. "Should no salt offer at St. Ubes you will then take up your Credit or carry your Cash to one or more of the Cape de Verdi, Touch at Bonna Vista, if you find none there, you may proceed to the Isle of May. Should you not meet with any Salt at this last place, after waiting 4 or 5 weeks then return to this Port with all possible Speed." These instructions for the Portuguese salt centers prepared by Hendricks stressed that no temptation should be yielded to in accepting freight of any kind other than the specified salt:

Should you be Boarded by Cruizers or Vessels of War belonging to any Beligerent powers of Europe your papers, being in such a state of regularity, no dispute can arise respecting the legality of your Voyage, according to the present regulations in Europe, as far as We have been able to ascertain: we wish you keep & look that the papers be in the State you now receive them; that in case of your being carried in, they may proove the innocence of the Voyage . . .

While the *Ophelia* was in port at Amsterdam the United States frigate *Chesapeake* was hailed off the Virginia coast by the British frigate *Leopard*. The *Chesapeake* was fired upon, boarded, and four of its seamen, alleged to be British subjects, were impressed into the English Navy. Up to this time, the British rules of search and impressment on the seas were restricted to merchant vessels, but now the first United States war vessel fell prey to the British. Hendricks' careful instructions had not been in vain. On July 26, 1807, two days after the *Ophelia* cleared Amsterdam for the neutral Portuguese islands, she was halted at sea and her empty hold searched "by his Brittanic majesty's ship Hyacinth" and she was "cautioned not to return to any Beligerent Port." The *Ophelia* luckily escaped the wrath of the British, but not without having her sea letter, which was authorized by Thomas Jefferson, docketed with the warning of John Davis, commander of the *Hyacinth*. The *Ophelia* pursued her course to the Portuguese coast, arriving at St. Ubes early in September. A cargo of 449 moys of salt was purchased with little trouble and, as soon as possible, Captain Waterman, obeying his orders, returned to New York.

Once the *Ophelia* unloaded her salt and was refitted with a new topsail of fresh ravens duck, purchased by Captain Waterman in Holland, she was ready for a fresh cargo. There was still time enough to load her hold and fling her sails with the last of the fall vessels. This time the *Ophelia* was bound for Cork on her way to Bristol. At Bristol she received a full tonnage of goods, and for the last time turned her prow in the direction of New York.

Meanwhile, in the closing weeks of 1807, Congress, smarting under the effects of the *Chesapeake* outrage, passed an embargo on American shipping that interdicted virtually all seaborne commerce, and ordered all British vessels of war to leave American territorial waters. In defiance of this order the British only increased their search of American merchant vessels and their impressment of American seamen. Caught in the turmoil that reigned on the seas, anxious to evade the French privateers, and to escape the marauding vessels of the British, the *Ophelia* was one of the casualties that followed the passage of the embargo. Back in New York she was reported overdue, and by May, 1808, Harmon Hendricks could no longer allay his fears for the ship's safety. Only the death of his two-year-old son, Joshua, reduced the loss of the *Ophelia* to insignificance. It was never determined whether the Yankee-built *Ophelia* was overpowered by the British and sunk, destroyed by the French, captured by Spanish pirates, or mysteriously swallowed up by the sea. Neither the *Ophelia* nor her faithful captain, Thadeus Waterman, was ever heard of again.

VII

The Embargo of 1807 was intended to coerce the English into withdrawing their policy of seizure and impressment. This, it was believed, could best be achieved by injuring the commerce of the English; but by eliminating American sources of supply the embargo proved harmful to the American shipping merchant. In 1807, 489 American ships cleared the port of Liverpool for New York City, but by the spring of 1808 most of these ships were rotting as they idled in the wharves. As a result the American shipping merchants sought ways of eluding the embargo. Smuggling was revived, subterfuges were used to enter forbidden ports, and trade was carried on over circuitous shipping lanes. English shipping to America was reduced considerably, but ships from Bristol, Liverpool, and London, although fewer in number, continued to arrive in New York, galling the American merchant with the sale of each shipload of merchandise. Widespread opposition to the embargo made unanimous support of Jefferson's policy impossible. The pressure of economic distress and its one-sided effect on the American shipping merchant, intensified the hostility to an act that was full of loopholes that favored England.

Whenever possible Hendricks urged the Bristol and Liverpool merchants to ship whatever was legally permissible under the act for his account to any eastern port; upon arrival it would be re-shipped to New York by a coastwise American vessel. At best these imports were limited. English firms became alarmed by the possibility that the Americans would be unable or unwilling to pay their bills. Freeman of Bristol assured Hendricks that his credit was excellent, and Hendricks, who often sent Freeman sums ranging from £300 to £1000 in advance of an order, responded by saying that "no act of my government shall or can prevent my paying you any Bal. Due."

The Hendricks correspondence with the English exporters during 1808 reveals less of a difference between the merchants than between their governments. English merchants described the conduct of their government as an intrusion of politics upon their trade, and most of the New York merchants took the same view toward Congress. Reaction to the embargo varied in other parts of the country. The New Yorkers joined the New Englanders in opposing the embargo, but Hendricks still withheld his personal views about the effectiveness of an act that failed to consider American dependence upon Britain and could eventually destroy the vital copper trade. As a matter of need he was compelled to turn to John Freeman when the city of New York, in the summer of 1808, made another of its many attempts to improve its water-works. The Manhattan Corporation engaged Hendricks to supply the city with the necessary copper bottoms that could only be manufactured in England because of their immense size. Otherwise, the summer passed by uneventfully; only apprehension about the future of the Atlantic trade continued.

Rumors of an amicable settlement of the differences between England and America became as popular as the belief that a war with France was imminent. The American merchant, doubting the wisdom of a Congress that busied itself with inconclusive debate, convinced that the carrying trade was unrelated to England's determination to control the seas, also believed that his English colleagues were subject to similar caprices of Parliament.

The embargo-induced depression in the shipping trade merely increased the use of copper. Hendricks reported to Freeman:

the sales I have made both in the quantities & at good prices exceed my best calculations; for the last two months the worms have so unmercifully pav'd away under ships, lying at our wharves, that altho' the times were so gloomy the merchants having good vessels, were compelled to shelter the Bottoms on copper, and were it not for the large quantity I had formerly contracted to receive from London I should have been out of copper . . .

For copper there was no gloom, and again Hendricks repeated his need to Freeman for large quantities. In 1804 Hendricks' requirements for his sales and factory were estimated at 200 tons annually. Four years later, when copper could be shipped only on English bottoms, he was anxious to purchase such a quantity from Freeman alone. To induce Freeman to give him preference over lesser customers Hendricks offered to pay one-third of the cost of an order in advance, and the balance thirty days after the arrival of each shipment. This was tempting to any merchant at a time when payment of a note within sixty or ninety days was equivalent to cash, but Freeman did not accept the proposal. In December of 1808 the situation was reversed when the breach between England and America was widened, and Hendricks informed Freeman:

. . . our political hemisphere has a cloudy appearance, the public prints will enable you to form a judgment on the various impediments to an amicable settlement with our European Brethren, and of the necessity of further restrictions being put on trade and intercourse. From the present complection of the times it appears to be imprudent to hazard further shipment on my acc.^t and risk on board British or foreign vessels to these United States but if your exertions are successful (and of your making the attempt I have no doubt) to give me a farther supply—it will much oblige me. Circumstanced as the two countries are I withdraw for the present any pretentions to further allowances, and at a more favorable opportunity will explain why . . .

In January, 1809, Hendricks joined the New York critics in lamenting the inertia of Congress. "Our Congress wasting time in debate," he declared at one time, and again to Freeman he stated, "the more our Congress proceed in debate the difficulty encreases to decypher the true intention towards the two hostile powers, things cannot remain in this state long . . ." With amazing foresight Hendricks placed extensive advance orders with the English merchants in preparation for any legal changes that might release a quantity of metals. Even then the fear of war cautioned Hendricks not to speculate on any doubtful items.

With Philip George, Jr., he increased his orders for square iron wire such as that used by the umbrella makers. George, always eager to accommodate Hendricks, replied that, "Unless you can ship the wire on board an American vessel insur'd, it will be imprudent to make a shipment as Politics have now turned." As a matter of fact, one correspondent informed Hendricks that some ship captains refused to accept mail, let alone take on freight. To complicate the export of iron wire, which was permitted, underwriters were reluctant to accept insurance for an article that was easily damaged by rust. With the Non-Intercourse Act in effect, this reluctance grew.

Using the information he obtained from George, Hendricks advised the Bristol firm of Harford, Partridge to ship his tin-plate by an American ship only. They, in turn, disturbed by shipping conditions that were becoming more and more irregular, lamented:

no American ship offers in this port yet on freight. One called the Dean has been loaded on the owners account, by which of course we had no opportunity to put anything on board, & fear there will be no vessel unless a very late one, or an English bottom, & owing to very rough weather the Coasting trade has been so much impeded that we have not had any opportunity to send but a small quantity to Liverpool, we therefore feel at a loss how to act with thy orders which are so positive, by an American ship only. . . .

George, despite his statement to Hendricks, risked shipping the iron wire on an English vessel. Harford, Partridge, adhering to their strict Quaker principles, or perhaps concerned with the difficulty in obtaining insurance for tin-plate, withheld shipment. Peter Maze, another of Hendricks' English metal suppliers, hoped for a speedy reversal of events, and John Freeman and Company patiently bided their time.

Meanwhile the American Secretary of State, Robert Smith, had entered into discussion with the British envoy, David M. Erskine.

In their exchange Erskine promised that the Orders in Council would be withdrawn early in June, and Smith assured him that American trade with Britain would be resumed. News of the agreement reached London early in May. While the English Cabinet discussed the merits of the Erskine-Smith proposals Peter Maze advised Hendricks that the Orders in Council had been withdrawn, although some of the blockading legislation was to remain. Normal trade was again to resume. Elated by this favorable report Hendricks informed the English merchants that the American restrictions would be lifted on June 10, 1809. American ships were quickly fitted up, loaded with goods for foreign ports, and with the Orders out of the way, they set sail amid the ringing of church bells and the booming of cannons.

Before the American vessels had crossed the Atlantic William Savill was writing from London what Harmon Hendricks had already read in the New York press:

You will have learnt from the public papers that our government has refused to ratify Mr. Erskines Engagements but that a temporary provision respecting the Orders in Council has been made for the relief of those American merchants who may have been induced to renew their intercourse under the supposition that the differences were finally adjusted.

The disavowal of Erskine's agreement as unauthorized added to the frustration of the New York merchant, who was again led to believe that the problems affecting his trade had been settled. Disconcerted English metal men, well stocked with goods, eager to sell to America, were embittered by the negotiations that failed to put an end to the differences between the two nations. Both reactions only increased the hostility of the merchants to their respective governments. The opportunity for an agreement between Britain and America had been lost.

During the short-lived respite that permitted American vessels to complete their voyages the Hendricks warehouse was replenished. If Hendricks gave the impression that he was short of goods, his inventory told another story. More than 11,000 boxes of tin-plate were on hand, a supply adequate for two years of Yankee peddling. With shipping temporarily open, he placed a tin-plate order that could accommodate him for two more years, and another order was negotiated for more than 20,000 copper bolt rods and approximately 5,000 sheets of copper. A previous shipment of copper, believed to have been sent before the revocation of the law, was seized as contraband. Hendricks, unaware of the cause for the seizure, prepared to take a loss of £1,500 but, without explanation, the copper was legally released. Further hope for adjudication of Anglo-American differences faded. Misunderstanding was succeeded by distrust. England stubbornly continued the search of neutral vessels and the impressment of American seamen. Humiliated but defiant, the Americans were unable to cope with the master of the seas.

The Non-Intercourse Act may have shaken the English economy but it did not budge her political outlook. Instead it throttled the American carrying trade and sowed harsh resentment among many of the shipping merchants toward the administration. The only shipments that were permitted to enter American ports were those negotiated prior to the Non-Intercourse Act. Only one copper cargo worthy of mention reached Hendricks at this time, coming by way of Puerto Rico to Philadelphia and, from there, on a coastwise vessel to New York. For the remainder of 1809 Hendricks' orders were curtailed. The new Non-Intercourse Act that followed the Erskine-Smith episode was strictly enforced. Even the packet boats carrying mail between New York and Bristol were spaced for a length of six to eight weeks, reducing communication and making it highly irregular. However the damage wrought by the impressment rules and the search of American vessels went beyond the damage done to commerce and trade. Hendricks was provoked to comment that more than commerce was at stake: "the respect due to our flag," he wrote with a feeling of national pride, was a matter of equal importance.

In spite of all the risks at sea the losses caused by spoilage, the high cost of marine insurance, and the acts that followed the Orders in Council since 1798 restricting trade across the Atlantic, Hendricks had benefited enormously from the profitable foreign commerce. A gradual transition from a wide variety of goods to an exclusive emphasis on metals enabled him to make copper and other nonferrous metals his specialty in the carrying trade. By the time that American relations with England reached a low point Hendricks was the largest importer of copper, brass, tin, lead, spelter, and iron wire in New York City. In his own state and the state of Connecticut he had no rival in tin-plate, and in the country at large he was known as one of the most extensive metal importers. When Hendricks dropped his usual restraint in asking Peter Maze of Bristol to ship him an additional five thousand boxes of tin-plates, he revealed the extent of his domestic market. "I have unquestionably in regular, not in speculative times, the largest and most numerous set of customers of any one man in the U. States." For the years before the War of 1812 no other similar body of records is available for comparative appraisal of Hendricks' claim, but the hundreds and hundreds of meticulously entered accounts in the Hendricks ledgers that have survived for this period easily attest to an extensive retail and wholesale trade. These accounts ranged from the obscure village tinsmith and handicraftsman in the interior of the country to the United States Navy.

The tin trade, less urban in its distribution than copper, was a secondary aspect of Hendricks' business. Prior to the War of 1812 it was he who set the pace of activity of the American tin market and, at times, controlled its price. Tin and tin-plate, "bright and shiny as a silver dollar," were, like copper, a major import from England. Their wide use gave employment to countless men, and the Connecticut Yankee who peddled the finished product up and down the Atlantic seaboard became the romantic symbol of the tin trade. Hendricks' tin competed with the product of the Connecticut smiths, and seldom did his stock fall below a year's supply. Eventually the widespread hoarding of tin induced by the Non-Intercourse Act glutted the market. Once the law was repealed Hendricks found himself the owner of a four-year supply instead of the two-year stock he thought he owned. Overnight the price of tin tumbled. The slump also brought about a decline in the price of copper, and speculation in metals of all kinds became rampant. Undisturbed by the new circumstance, Hendricks, instead of disposing of his tin stock, bought as much as he could of the hoarded tin from his competitors, who were forced by the inactive market to sell at any price. To friends in Bristol he wrote, "It is expected the Non-intercourse will this day end, and trade resume its former splendor, but sad will be the fate of some thousands of Boxes Tin'd plates, that will be throne in the market by unfortunate speculators at high prices." Hendricks' vast tin stock now enabled him to reestablish the former price of tin, but not without making heavy demands on his cash resources. More and more he was compelled to turn to the West Indian financier Jacob Levy, Jr., for large but short-term cash loans.

The pause in trade brought about by the temporarily depressed metal market encouraged Hendricks to rely more on the importation of Russia goods, woolens, and leathers. He undertook to fill orders of a distinctive sort for special customers when copper was not available. Surgical and dental instruments, fine tools, choice glassware, French laces, and English piping—quality articles that were not manufactured in the United States—appear in the orders of 1810, when the importation of tin was avoided. Greater attention was devoted to the domestic and inland trade, to the needs of the shipyards, to banking, and to the acquisition of real estate. Meanwhile Hendricks continued to hope that the differences with England would be settled.

VIII

During his years of trading and importing, other business interests had gradually drawn Hendricks' attention. Since 1799, when he bought his first hundred shares of Manhattan Bank stock, his activities were intertwined with every phase of early-nineteenth-century investment. Investment was such an integral part of the career of an enterprising merchant that it would be difficult to reconstruct Hendricks' activities without an understanding of his combined role as merchant-investor.

Following his father's methods, Hendricks utilized the successful practices of the eighteenth-century merchant. These practices remained in vogue during the first decades of the new century, but Hendricks' ability to adapt them to his own needs set him apart from other young merchants. He replaced investments that were based on the system of sharing in joint ventures in the purchase and resale of sugar, or in a vessel of goods for the West

Indies, by investments in metals. Funds kept on deposit in St. Petersburg in the early 1800's for the purchase of Russia goods through an English agent were applied to the purchase of copper direct from the English rolling mill, without recourse to an agent. The use of surplus capital for purposes other than commodity transactions, such as the acquisition of government stock and bank shares, was one method of investment that had a functional relationship to the carrying trade. Another good area of investment was the purchase of real estate and advancing mortgage funds. These, too, were the outcome of business circumstances. A third was in the discounting of notes and, finally, Hendricks invested in canals, turnpikes, steamboat lines, and in heavy industry. Each aspect of investment was related in some manner to copper, the mainstream of Hendricks' activity.

Immediately after the purchase of the Louisiana Territory the United States issued stock in the amount of $11 million. Hendricks made extensive purchases of this 6 percent stock, to which he added purchases of United States Bank stock, and the issues of many state banks and insurance companies. Bank and government stocks—those whose interest was payable abroad—were applied to the payment of English bills when bills of exchange could not be transacted with ease. These stocks were also used as a fund for deposit with English firms and drawn upon in advance of a major order. A power of attorney accompanied the stocks, in order to make them negotiable at the proper time. This practice gave Hendricks an advantage when purchasing desirable goods and frequently reduced the problems associated with the remittance of funds.

Hendricks developed an interest in finance and acquired a knowledge of investment that ultimately brought him into banking. He did not follow the practice customary among other merchants of discounting notes in order to raise funds. Instead he sold most of his notes—those paid to him and those he bought at discount—to Jacob Levy, Jr. In one thirteen-month period $83,-000 in notes were transacted between the two men, and later these transactions soared to $250,000. Since Hendricks bought the notes at low rates of interest and resold them at a small margin of profit, he constantly had available the large sums of ready money that he required to purchase copper when the price was low, to face the crisis in the declining tin market, and to purchase bank shares and real estate.

From the time that Hendricks purchased his Greenwich Street summer house in 1801 as a refuge from the yellow fever he was attracted to real estate investment. It is not known whether the northern movement of the city's population influenced his venture or whether the gradual acquisition of property was at first a step in expending surplus money. His purchase of the Bedlow Street house in 1804 was motivated by manufacturing interests and for its use as a foundry. The house at 76 Broad Street was necessary for comfortable living quarters, and the lots and ground adjoining the Greenwich Street house were prized for their spaciousness and beauty. In 1807, when the properties at 118 and 119 Commerce Street, at the corner of Bedford, were purchased, and in 1808, when six acres of land along the Hudson were added to the list, it was with a view toward investment. In 1810, when a site was acquired at 37 Beaver Street in an industrial section, it was for the purpose of erecting a smelting house. The $34,000 worth of real estate accrued by Hendricks by 1810 was exceeded only by his stock and banking investments.

Once the Bedlow Street foundry was put into operation for the smelting of copper and the cutting of bolts, Hendricks' interest in manufacturing grew yearly. To obtain a better knowledge of the state of copper manufacture he had visited Revere's rolling mill at Canton and the site of the Schuyler works near Newark, which had been renamed the Soho Company. He advised Haym M. Salomon, who plunged into copper manufacturing, a business he did not understand, on the problems of a rolling mill. Salomon was eager to establish a small mill of his own in New York City but his venture into manufacture was brief and unsuccessful because of his inability to overcome difficulties arising from faulty equipment. Salomon turned to Hendricks, who advised him to consult with Revere. Revere was faced with the same technical problems, and he sought to overcome them by sending his son to England to obtain advice and information on the rolling of copper. Even Revere's thorough knowledge of casting bells could not eliminate his dependence upon English-cast rollers. With the experience that he had acquired over many years he was able to correct defects, but no one in the United States could produce the precision rollers required by copper mills. Unable to obtain proper equipment and lacking Revere's skill and experience, Salomon was forced to dispose of his small mill. Because their meager knowledge of metallurgy neither of the men realized that many of their difficulties with the rollers were caused by the impurities that were not removed from copper.

The Soho Company mill, operated by Roosevelt and his associates, possessed equipment superior to that owned by Paul Revere. They had the advantage of an adequate year-round water supply and a nearby copper mine, but they were unable to utilize these facilities because they were constantly in need of funds. Shortly after the Soho Company fulfilled its contract with the United States Navy the mill suspended operation.

After observing these difficulties and failures Hendricks set aside any plan which he might have had for a mill of his own. But his interest in manufacture did not decline; instead it became more evident after 1806 than before. His inadequate knowledge of technology made him more aware of the need to pay attention to mechanics, engineering, and the application of steam in industry.

Hendricks' recognition of the value of steam power made him an early advocate of its use. He encouraged young inventors such as Stedman Adams of Connecticut, who was seeking approval for his steam invention in England. Hendricks described the object of the invention to his cousin Michael Oppenheim of London and requested him to tender Adams every possible courtesy while in England:

Mr. Adams object to England at this time is to give circulation to a very valuable invention entirely of his own construction for working mills by Steam thereby supplying the want of watter falls for large and small manufacturers which may be increased or diminished according to the nature of the articles required to be produced from Mills, and to Suit small cappitols as well as large . . .

A greater opportunity to observe the uses of steam presented itself the following year, when a new era was opened in the development of American transportation, an era that also marked the beginning of a new phase of interest and activity in the career of Harmon Hendricks.

In the spring of 1807 the naval architects and shipbuilders Adam and Noah Brown purchased copper from Hendricks for purposes other than sheathing. Brown's shipyard was busy fitting up a boat to be driven by steam. Built under the direction of Robert Fulton, who had just returned from Paris, this boat was epoch-making in American maritime history. Robert Livingston, who had supported Roosevelt in copper manufacturing, had now joined with Fulton in an effort to build a steamboat that could navigate the Hudson River from New York to Albany. When the steamboat *Clermont* was launched, it was destined to become the first inland steam packet in the world. The $653.12 worth of copper supplied by Hendricks to the Browns at the time the boat was being built was possibly used for the boiler of the *Cler-*

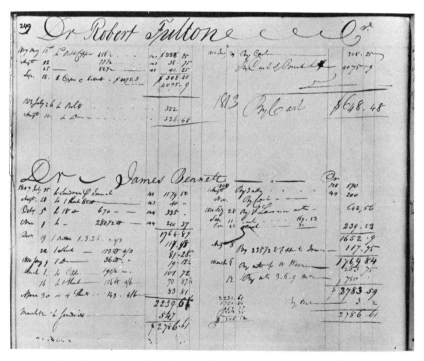

Partial account of Robert Fulton with Harmon Hendricks.

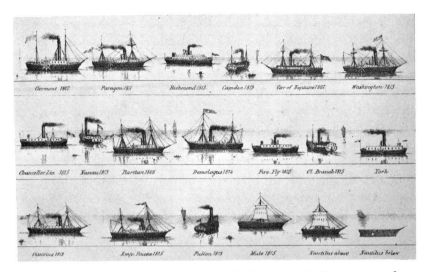

Robert Fulton's steamships, nine of which obtained their copper from Hendricks.

mont. From the time that Fulton returned to the United States until his death in 1815 much of the copper that went into his now famous boats came from Harmon Hendricks. It was not an uncommon sight for Fulton to visit the Hendricks warehouse, select his copper, and instruct the carters about hauling the copper bolts and sheets to the shipyard. Hendricks' contact with Fulton and the influence of the new uses for copper indicate that he was alert to every manufacturing innovation that related to the use of the rust-brown metal.

Furthermore, as Hendricks noted, interest in American manufactures was sharpened by the long embargo and the commercial restrictions. The reduction of the flow of imports was bound to affect the production of American goods. An obvious change was reflected in the conduct of the shipping merchant, who turned his thoughts to manufacture. All of the pamphleteering encouraging American manufacture was not as convincing as an open market that could not be supplied. Manufacturing enterprise gradually attracted the idle capital of the shipping merchant who was deprived of his former sources of supply. If Hendricks was an example of the shipping merchant who stepped out of his mercantilist role into that of industrial capitalist, then the beginning of this transformation can be traced to the five years preceding the war. His English colleagues also quickly became aware of these changing circumstances when they saw the products of their mills lying in their warehouses and stacked along the British docks. As the stockpiles grew, letters to Hendricks from England expressed a deep concern for the loss of the American market. Behind this was a fear that the lack of commerce would serve as an impetus to American manufacture, which in turn would be disastrous for the English mills.

In the summer of 1811 Hendricks balanced his doubts and hopes for a settlement of Anglo-American differences. He wrote to John Freeman about the changing conditions within the American metal trade:

Should our commercial relations assume a brighter appear[ce] I will again remit in hopes of better success than my last, and I trust you will excuse the trouble I have been [on] occasion to you arising from an anxious wish to faciliate an early supply of your article. Attempts will be made to manufacture Braziery Copper shortly. Sheathing copper has been attempted with much success and will be pursued by others. The monied capital of [the] U.S. being mostly unemployed in imports, many new manufactories will be set on foot to give employ to the poor as much as benefit the rich. Many articles of British manufactories bear an advance of 100 to 300 pr. ct. from last years price which holds out great inducements to supply domestic substitutes for British imports.

Hendricks' letter provoked the Freeman firm into inquiring further about American manufacturing plans. Allaying Freeman's fears that there was an immediate danger of American manufactures replacing English imports, Hendricks gave his evaluation of copper mining by assuring his correspondent:

Copper is not to be added thereto, labour to work the mines is yet to high priced in this Country, 2 to 4/ stg. per diem, being the average price of the most servile kind. I have resisted all the applications to me for that object from a persuasion of being unsuccessful.

The high cost of labor that slowed the development of copper mining did not affect other aspects of Hendricks' metal interests. Successful domestic innovation with specialties eliminated such imports as the finely hammered gold and silver leaf which was purchased in London. This English handicraft was improved upon in the United States when rollers were developed and introduced for its manufacture. Domestic substitutes, once they had been perfected and could meet the demands of the trade, became permanent. Wherever superior methods of production and timesaving devices could be introduced to overcome the labor shortage they were quickly adopted. Between 1810 and 1812 the importation of these desirable specialties was gradually eliminated by Hendricks when the new industry began to flourish in New York. The gold and silver leaf once supplied by Atkins and Harford, Partridge now disappeared from the Hendricks' ship manifests. Although technological developments were slow as yet and hardly of great importance, the door to American industry had been pried open by the high cost of goods and the decline of imported manufactures.

The American merchant was still too timid to invest his capital in domestic enterprises. He continued to look to Congress in the hope that some agreement would reopen normal trade. But this hope grew fainter as English-American relations worsened. England retaliated against the American tactic of peaceful coercion by refusing to grant export licenses and maintaining her blockade. Offers by English manufacturers to Americans to evade the English blockade by shipping goods to Spanish ports for reshipment to the United States were rejected; orders that were six months or a year old remained unfilled. The once-thriving carrying trade was brought to the edge of ruin. Only a fraction of the millions of dollars in goods that had been imported annually by the United States now entered her ports. Business failures in England increased daily; British manufacturers and merchants were driven to despair as they, too, anxiously awaited an amicable settlement. Failures in England caused bankruptcies in the United States when American funds on deposit were lost in the crumbling British firms. In the United States the state of commerce was equally discouraging, and the New York merchants who had steadfastly opposed the non-importation measures fully felt their effects. Merchants on both sides of the Atlantic wrote about the inability of their governments to come to any agreement.

To avoid endangering his English funds Hendricks requested his trusted colleagues in London and Bristol to invest his advance remittances for unshipped copper in United States government stock. This assured some safety for his English fund, prevented it from lying idle, and guaranteed ready cash for copper when it was again available for shipment.

Merchants on both sides of the Atlantic were again led to believe that trade would be resumed when arguments in Congress favored the admission of copper orders, negotiated in the fall of 1810, to American ports. And in England Parliament, under the stress of an injured economy, again reviewed the Orders in Council in the belief that an easing of the restrictions would reduce her well-stocked warehouses which contained goods intended for America. Again the conduct of both governments confused the merchants and revived a false hope in the renewal of trade. At this time Paul Revere, together with Hendricks, ordered 4,000 sheets of Bristol copper, presumably for the United States Navy. Meanwhile American naval purchasing agents, fearing a continued shortage of copper, quietly purchased all available ship copper in New York, thereby considerably reducing Hendricks' choice stock.

The promise of trade that appeared during January and February of 1812 was no longer believed by March. Hendricks, again uncertain of the possibilities of purchasing copper, turned to John Freeman for a quantity of sheet iron. This could be bought in Russia and shipped on one of the American vessels that reached the Baltic ports. The purchasing power of Hendricks' English funds, formerly applied to such transactions, was now of little use. He arranged to have £3000 of uninvested money applied to the purchase of 100 shares of 6 percent United States bank stock, thus protecting the balance of his English funds against the now imminent dangers of war.

In April the debates in Congress, hovering over peace or war, offered little promise to the American merchant. The New Yorkers publicly expressed their dissatisfaction with the protracted congressional seesawing that recommended still another embargo. Through the window of his counting house Hendricks could see the throng of merchants in "great confusion & uproar in the anticipation of an Embargo . . ." English and American merchants continued to lament the inability of their governments to reach an understanding. "Calculations cannot be made on Politics," wrote Hendricks to Pieschell & Schreiber. "I have none to hazard but must await events good or bad with patience."

The doubts, the promises, and the speculations about the immediate future of trade soon came to an end. English impressment rules were unaffected by Jefferson's embargo; non-importation failed to persuade Parliament to rescind the Orders in Council that restricted her merchants from exporting to the United States, and no act of Congress was able to drive English vessels away from American waters. Although a politically divided Congress, representing a variety of undefined mercantile and sectional interests, wavered between making its decision for peace or for war, many of its members no longer restrained their cry for forceful action. When Congress finally mustered the strength for a decisive vote the anti-war men were outnumbered. On June 19 James Madison, despite opposition from his own party, and from the Federalists, declared war against Great Britain and her dependencies.

The merchants and shipowners who had suffered the effects of impressment and who had been stung repeatedly by the restrictive Orders in Council saw little benefit in a war that would completely deprive them of the once-rich Atlantic trade. The English smelters and copper manufacturers were genuinely alarmed by the knowledge that Americans were prepared to venture into copper manufacture. Non-importation had provided the first stimulus to American manufacture; a war could complete the task. However, American copper importers were still hesitant about the problems of manufacture.

Metal merchants were divided in their opinions of the war. Hendricks had spoken before of the humiliation of American ships by the British, and to his English correspondents he had expressed regret about the lack of respect shown for the American flag. With little elaboration, but with the same concern that he had previously shown, Hendricks summed up the situation simply: "the unhappy differ[enc]es between our governments, has terminated in war. . . ."

4

THE SOHO COPPER WORKS

The Rise of an American Industry

I

In the fall of 1812 Harmon Hendricks was able to relax fully for the first time in more than a decade, and he exchanged the world of copper for a long vacation. A trip with his family through the interior of New York State provided the relaxation that he had sought and, after an absence of five weeks he returned to a business that had come to a standstill.

The New York to which the Hendricks family returned was seething with political discontent. But like so many other New Yorkers of their milieu they were as yet unaffected by the conditions of the war, which was only a few months old. Hendricks, however, had read with pride of the skill of American seamanship that had been demonstrated in the engagement between the United States frigate *Constitution* and the British frigate *Guerrière*. He reread the newspaper accounts of this outstanding American naval victory with considerable satisfaction and, like so many merchants interested in naval affairs, he looked forward to a victory on the seas. Like most New Yorkers the Hendrickses eagerly awaited the war news that was belatedly reported in the New York press.

The Hendricks family of six children was a lively group. Hetty, the oldest daughter, busied herself sewing samplers, practicing her dancing routine, and studying geography. Energetic, ten-year-old Uriah studiously followed his Latin and Greek master. His little brother Henry already looked forward to the day when he could exchange his ciphering books for the stories of Caesar and Homer.

Hendricks now had the leisure to share in the newly acquired knowledge of his children, and found time to join them in their French and Spanish recitations. The knowledge of French that he had acquired through study with the aid of his brother-in-law Abraham Gomez was useful, although previously he had used it primarily to read Gomez's letters from Bordeaux. Now it was a great source of pleasure to speak French with the children. In the

Uriah Hendricks' eighteenth-century register of births and deaths.

huge copper ledger he kept a family register of births and deaths in his best French, a contrast to the neat, square Hebrew characters in his father's hand which recorded the history of the family during the previous generation.

Simeon Levy, whom Hendricks had befriended, was a regular visitor in his capacity as a teacher of Hebrew and the Bible. Hendricks did not share the view of his West Indian friends about Hebrew. He would not deny his children an education in the language of the Bible in exchange for French and Spanish merely because those tongues were so necessary to the world of commerce. He gave them both.

Among other visitors who brought pleasant days to the children was Uncle Sol Isaacs, whose gift of a new toy or a soft leather doll for one of the girls, or a freshly bound copy of the popular *Columbian Orator* for Uriah made his appearance a welcome one. A visit from their Charleston cousin Mordecai De Leon was also a happy occasion. Mordecai, following the path of his brother Abraham in the study of medicine, would occasionally come up from Philadelphia where he was attending medical school. Hendricks had no desire to encourage his sons to enter the medical profession. He was eager to provide them with the advantages of a general education and then apprentice them to the copper trade. For the first time Hendricks had begun to show an interest in the future careers of his young sons.

Before the war was declared two consignments of copper left England and unexpectedly reached Hendricks while he was contemplating the future of business. One small copper cargo on the ship *Harriet* was seized by American revenue officers as it passed New Haven, but it was released and permitted to enter New York harbor. Revere shared an interest in this copper and was informed of its safe arrival by Isaacs. A second shipment, on the *Independence,* which comprised copper from Bristol, Liverpool, and London, was taken in by American privateers in the fall of 1812. The United States marshal disputed the privateers' claim and permitted the entire cargo to be released to Hendricks. These two shipments were the last that Hendricks received, and what benefit he obtained from them was lost a few months later when he attempted to run the English blockade with a shipment of pig copper to France in exchange for a quantity of silks. The effectiveness of the British blockade, which covered most of the major eastern ports, was felt along the entire length of the Atlantic coast.

Hendricks scrutinized the ship arrivals from South America for unconsigned shipments of copper, and saw the difficulties shipowners had in selling undesirable Russian pig copper at the price of 30¢ per pound. He read the published listings of prizes and noted that the privateers who required copper for their own ships returned with little or no metals. Captured prizes were now subject to a host of charges by the government and were frequently more profitable to the government than to the privateer who risked his life in capturing ship and cargo. Hendricks had no interest in privateering and, as he had in the past, he continued to view with disfavor the professional smugglers who daringly entered forbidden ports in search of goods that could be sold in the United States. But many of the shipowners and shipbuilders envisioned fat profits by outfitting their idle vessels to prey on enemy shipping.

While a final attempt to seek peace between the British and Americans failed, the once-copious correspondence between Hendricks and the metal merchants of England thinned down to a single letter. Hendricks estimated the amount of his undelivered copper that lay on the docks and in the warehouses of Bristol and Liverpool as he watched copper in the United States triple in price, going from 20¢ to 62½¢, then rise to 78¢ and drop to 50¢, before it soared to the enormously inflated price of 80¢ per pound. In the first months of the war, when the British blockade closed Chesapeake and Delaware bays and correspondence between the two countries officially ceased, Hendricks realized that he had no choice but to temporarily forget his frozen English assets in copper and bank stocks.

In this reflective mood Hendricks watched the disgruntled American shipping merchant, who had built a small empire out of the carrying trade, search for other avenues of business. Some of the merchants showed interest in the long-talked-about program of developing American manufacture, others looked to the sale of war materials or privateering, but all were faced with the problem of compensating for the loss of the carrying trade. An era had come to an end. Hendricks, still reflective, withdrew from his customary activity. Life at the Greenwich Street house continued as usual, with only the newspaper accounts of the war penetrating Hendricks' intimate family circle.

The collapse of the shipping trade and the depressed business conditions that followed the declaration of war failed to alarm Hendricks. To the contrary, his confidence in the government was evident when he subscribed $16,000 to the $11 million government loan of July, 1812. In doing so Hendricks again showed himself to be different from the other New York merchants, many of whom opposed the war. Although many of the merchants who had been enriched by the carrying trade expressed the fear that the new conditions might make them destitute, Hendricks observed that those who had emerged as wealthy men could console themselves in the knowledge that a comfortable living could be made from the interest on the monies they had amassed. The ordinary channels of investment associated with the carrying trade were swept away. Hendricks' new role under the changed conditions was not yet clear. But the months that lay ahead indicated that he accepted his own counsel that a comfortable living could be earned from interest on his funds, and he utilized this source of income by investing in stocks, banking, insurance companies, and real estate. This trend was becoming prevalent in the mercantilist circles of New York.

Hendricks' correspondence with James Ward, which had once concerned itself exclusively with copper, tin, and personal gossip, now centered around the subject of Ward's banking interests. It was at the insistence of Ward that Hendricks purchased 300 shares of the new stock of the Middleton Bank in Connecticut. This was quickly followed by sizable purchases of stock of the Pennsylvania Company in Philadelphia, and the American and Mechanics Bank of New York City. Again at the prodding of Ward he subscribed more than $61,000 to the Hartford Bank, of which Ward was a director. Through Ward Hendricks also became a director of the Hartford Bank. Three months later, in the spring of 1813, when the full force of the British blockade was felt at New York harbor, Hendricks subscribed $30,000 to the new loan of the Manhattan Bank, in which he was an original shareholder. He also subscribed an additional $20,000 to the Mechanics Bank. In much less than a year Hendricks' bank subscriptions exceeded what was then the fabulous sum of $100,000. Meanwhile the Americans had been forced by the British to fall back at Detroit, spreading fear among the garrisons of that area. Hendricks, who was well informed of all war news, was now placed in the strategic position of counselor to the cashiers and officers of the banks with which he was associated, because of his good judgment in financial matters. Yet even this fruitful branch of finance did not distract him from his interest in utilizing his knowledge of metals and searching for a copper supply.

The doldrums that characterized the copper trade provided Hendricks with more leisure for his small but substantial library. Banking hardly consumed his time and copper sales were at an all-time low, so there was ample time to visit the New York Society Library or call upon his brother-in-law Benjamin Gomez, the bookseller. On his way to the docks Hendricks could see the newly outfitted privateers ready to set sail, although the enemy craft were close enough to be seen on the distant horizon on any clear day. Gomez might supply Hendricks with a book on astronomy, one on geography, or perhaps one on the sea. Hendricks' taste for the natural sciences was exceeded only by his desire to obtain a better knowledge of the application of metals to industry. Gradually he acquired a library dealing with the arts of manufacture. The standard encyclopedias that lined his shelves were typical of those used by the early-nineteenth-century merchants, and they were flanked by volumes on mechanics and engineering. Mining tracts from England were heaped together with American pamphlets on canals, turnpikes, navigation companies, and banking. Books on science and trade stood alongside the mercantile directories of major American cities. Circulars detailing European and American prices current of preceding years lay unused. There was a notable absence of ready reckoners and standard interest tables, for which Hendricks justifiably professed no use.

Of the dozens of pamphlets that were ever present on his orderly desk there was one that Hendricks had obtained in 1801 while he was still located on Pearl Street. It recounted the story of the Schuyler copper mine up to the time when it was reorganized under the name of the Soho Company. Hendricks had read it, studied it, and reread it. He began seriously to think in terms of buying the Soho mine and mill. He had visited the mill shortly after Roosevelt had reorganized it, and he had revealed some concern about its potential production, but he was strongly impressed by the men who ran it: Josiah Hornblower, who planned its engineering facilities; Nicholas J. Roosevelt, who envisioned substantial profits from operating it; and Chancellor Robert Livingston, who helped finance it.

In 1813, when Hendricks decided to attempt to alleviate the crisis that faced the copper trade, his interest in Soho was renewed. He was not the only one to hold an interest in copper manufacture. Robert Fulton, who complained of the brittle copper produced by Revere, also considered the possibility of manufacturing copper in the New York area, but his preoccupation with steamboats prevented him from doing so. The lack of manufactured copper could have disastrous results upon the trade, and Hendricks, who possessed the courage and vision required to invest in a productive and important manufacturing enterprise with which he was familiar, took the necessary steps to do so.

Hendricks' decision to buy the Soho Company, made within a year after the war had begun, was the first step in his transformation from a merchant-investor to an industrialist. In this remarkably short time Hendricks had determined on a course of action that was to influence the lives and fortunes of generations of Hendrickses to come. The decision would also play a key part in the development of the metal industry in the United States, and was to establish the name of Hendricks as a leader in American copper for the next century.

The Soho works, soon to become a central force in the history of the Hendricks family, took its name from the famous Bolton and Watt factory in Birmingham, England; it was described as the best-equipped engineering plant in the United States. It has not been noted previously that Soho was operating two blast furnaces for smelting and refining copper before Paul Revere opened his copper mill at Canton. Its proximity to the stream flowing from the Passaic River gave the mill an ample year-round water supply. As a pioneer copper mill in America Soho had re-

The old water wheel at Belleville, New Jersey, on the site where the engine was built for the first steamboat.

ceived one of the first known copper contracts from the government of the United States, had built the machinery for the Schuylkill Water Works of Philadelphia in 1800 under the direction of Benjamin Henry Latrobe, and had engaged in similar undertakings in its brief history of less than a decade. Why the Soho Company did not succeed has not been determined. Its mines continued to yield ore and its mills produced American copper as late as 1806. When the city of New York required copper to improve its waterworks the machinery at Soho was already silent, and the Manhattan waterworks requested Hendricks to order the necessary materials from England. By this time the fate of Soho seemed certain. Roosevelt caught the steamboat fever and left New Jersey in 1809 to try his luck with steam navigation on the western waters. Livingston had previously become the American minister to France, where he met Robert Fulton, to whom he now gave his financial support. And Hornblower, an ailing man of seventy-eight, was destined to live only one more year. The seat at Soho

Ruins along the Second River, Belleville, New Jersey.

was deserted by 1808, and its stockpile of American copper remained untouched for the next five years while American merchants searched for copper from abroad. Its fine machinery became an undisturbed refuge for the spiders, who covered it with their artfully spun webs. Rubbish choked the drifts above the mine shafts that had been drained of water to make the mine workable, and the only mill that operated in the area was one recently built that ginned cotton. Not far from the cotton mill was the Belleville Gun Powder Works, which was active in producing ammunition for the government, but copper at Soho was forgotten. For five years the Soho works stood inactive, a quiet testimony to the failure of one branch of American industry. Some of its equipment was stripped, and the mill became a barren sight amid its twenty acres of surrounding land, a quarry, and a brook known as the Second River.

To reactivate the deserted mill and use its facilities for reconverting old and raw copper Hendricks had first to assure himself that he could overcome the difficulty of obtaining experienced

Water wheel at the ruins near Hendricks Pond, Belleville, New Jersey.

mechanics, meeting the high cost of labor, and finding a man qualified to operate the mill. He chose Solomon I. Isaacs as the mill overseer. Isaacs, raised to the copper trade, was a natural choice. He understood the methods of refining and rolling copper and was familiar with every aspect of copper from the time it was mined to the time it was nailed to the hull of a ship. He was the most perfectly suited person to tackle the problem of bringing together a force of trained mechanics and skilled mill hands, at a time when the conditions of war had absorbed most of the trained men in the New York area. At the age of twenty-nine, Isaacs was probably the best informed person on brazier copper in New York City. In choosing Isaacs Hendricks solved a problem that might have stymied other entrepreneurs.

Once Hendricks had decided on Isaacs he opened negotiations for the purchase of the Soho mill and the estate at Belleville, New Jersey. This gave him the distinction of being the first copper merchant in the United States to step outside of the mercantile role to explore the path of industry.

The Soho tract, with all of its improvements, was appraised by
Roosevelt in 1801 for $65,000. When negotiations for the mill
were completed on May 1, 1814, Hendricks purchased it for
slightly less than $41,000. Isaacs, the mill overseer, was admitted
to terms of equal partnership, and though he advanced no funds
of his own he compensated by his knowledge and experience.
Hendricks' new firm adopted the name of Solomon I. Isaacs, and
Soho Copper Works. Isaacs, in addition to being overseer, was
also given full authority over the mill at Belleville. It was also
agreed that the contract between the two men was to be for a
four-year term and was to operate independently of any other
branch of Hendricks' business.

Plans for the construction of a new smelting house and a fully
equipped rolling mill were immediately put into operation. To
facilitate ready transportation a private road was built from the
mill to the Second River, and to further guarantee access to the

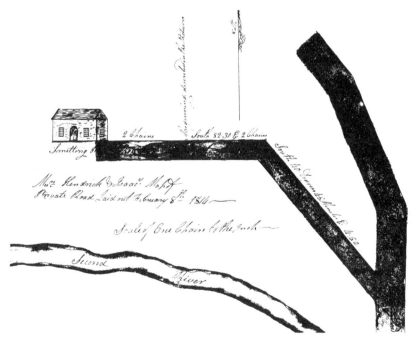

Section of water-color map of 1814 showing Hendricks' property at Belleville.

mill road twenty shares of stock were purchased by Soho in the United Hackensack and Passaic Bridge Company. Forty shares were bought of Elisha Boudinot's newly organized York and Jersey Steam Boat Company, which would eventually ferry the manufactured copper down the Passaic River. All shares of the bridge and steamboat ferry stock were divided equally between Hendricks and Isaacs, Hendricks assigning his shares to his young sons as an investment in their names, and as an inducement for their future entry into the business. The New York warehouse on Beaver Street was to have its own smelting house and office. With the purchase of the bridge, transportation stock, and the horses and carriages, the total expenditure reached $122,305. This tremendous investment in a new industry at once outstripped all efforts made by Revere, who claimed that his investment of $25,000 in the Canton rolling mill had been jeopardized by a lack of encouragement.

It was with considerable pride that Hendricks later entered this revealing note in his ledger:

The Soho Company commenced Business in May 1814—without one Cent of Capital, enter'd soley on Money Borrowed at 7 pr. Ct. pr ann.

At no time did Hendricks provide an explanation for the loan used to purchase Soho. The considerable extent of his other investments hardly disturbed his resources or compelled him to seek funds elsewhere, for at this time he was involved in a number of other transactions. In the month of May, 1814, while the British, who had just overthrown Napoleon, were preparing to concentrate their strength in a vigorous attack upon the Americans, the mill partnership was established, and Hendricks completed the purchase of Hannah Jacob's Greenwich Street house for $17,000. On May 20, the day he subscribed an additional $42,000 dollars to the desperately needed $20 million war loan, he also bought 20 shares of stock in Fulton and Livingston's newly organized North River Steam Boat Company, paying in cash. In addition to this he acquired 50 shares of stock in a patent cloth factory. Apart from the Soho mill, his total investments reached approximately $250,000 and, by comparison, the copper that was being prepared at Beaver Street for rolling in New Jersey seemed insignificant. The new trend in the American economy, investment in steam, in cotton mills, and in small factories, the real genesis of American manufacture, caught the imagination of Hendricks. Without the full application of his business acumen at this time the manufacture of copper could not have been advanced. Though he never surrendered his interest in the carrying trade and he retained his mercantilist position, neither did he lose his interest in copper, and now he fused his talent and energy into a plan for its manufacture.

If the four-year clause in the partnership agreement between Hendricks and Isaacs was an indication of their idea of the expected length of the war, then Hendricks felt safe in making a wager with Jack Cohen that hostilities would not have ended in four months. At the end of four months Hendricks won the bet and a good beaver hat.

The bet may have been more serious than a spur-of-the-moment wager, for Hendricks had subscribed a total of $58,000 in government loans when money was sorely needed and merchants were unresponsive to President Madison's call for financial support. Hendricks' knowledge of British superiority and his national pride compelled him to respond more quickly and more liberally than others. At the end of this four-month period, however, there was little doubt in anyone's mind that the war would last for an undetermined length of time.

The family's personal involvement in the war picture came when news arrived from the Charleston De Leons in October, 1814, that Mordecai Gomez Wagg had died. Wagg had joined the frontier forces of Andrew Jackson, was commissioned a major, and "his tour of duty was through the Indian Nation, when on his return through Savannah, Georgia after an absence from us of eight months, Death put a stop to his honourable career and at once blasted the high raised expectations of his family and Friends." The unpleasant task of informing the English cousins that their brother was a casualty of the war fell to Hendricks.

Before the war was over Soho was completely refurbished. Motive power was obtained from the Second River. The smelting house was rewalled with new fire bricks; rollers that could produce sheet copper in various thicknesses were installed, and the stamping and cutting machinery for plates, nails, and spikes was set into operation. More than 2,300 pigs of copper were sent down from Beaver Street to the mill, and in October of 1814, almost on the day that the British left Chesapeake Bay to sail for Jamaica, the first sheets of copper were rolled at Soho. Seixas in Philadelphia received the first consignment, which was quickly sold; and shortly thereafter, the products of the mill were advertised in the New York press:

The Soho Copper Company, will have finished in a few days prior orders for Sheet Copper and Bottoms; are ready to receive further orders to manufacture Sheathing Copper 16 to 36 ounces. Braziers and Steam Boiler Copper 12 to 56 inches wide, 4 lbs to 250 lbs per sheet. They have now for sale at their warehouse, No. 37 Beaver Street, a few Sheets of various sizes and weights. Also Calk or pickle dust for Oil of Vitriol, Copper Sheathing, Nails, Rivets, Still Cocks & Block Tin. Letters postpaid will be attended to, addressed to,

S. I. Isaacs, and Soho Copper, New York

Among the master shipwrights who went to work at Lake Erie for the American Navy, under the direction of Commander Oliver Hazard Perry, were Hendricks' associates in the shipbuilding industry, Noah Brown, Christian Bergh, and Henry Eckford. The shipbuilders worked faithfully with Perry to build a navy that could stand up to the British. More than eight hundred ship carpenters, caulkers, metalsmiths, and laborers were drawn to the Lakes from the New York City area, and the shipyards that once employed them were now deserted. Only Adam Brown attempted to keep the Brown Brothers' yard open. Here, in the spring of 1814, while Hendricks was busy rehabilitating Soho, the Navy gave its sanction to Robert Fulton to build the first steam war vessel, the *Demologus*. There is no record of how much of the copper Brown purchased from Hendricks went into the *Demologus*, but it is likely that all of the nails and sheets he supplied were used in the pioneer war vessel, for no other ship was being built at this time. Perry, meanwhile, won a decisive victory over the British on Lake Erie. The ship workers, their task completed, gradually returned to New York. The yards were reopened one by one, and while Soho embarked upon the production of war-time copper the welcome news of peace reached New York.

Although young Oliver Hazard Perry emerged as a great naval hero as a result of his triumph on the Great Lakes, this victory was not proof of American naval power. The war had, in fact, clearly demonstrated the vast superiority of British seapower. The English fleet had maintained an effective blockade of the major ports along the Atlantic coast for the duration of the war; the city of Washington was captured and burned in the summer of 1814, Baltimore soon after, and Philadelphia also appeared to be in danger at this time; and in the waters outside of New York, the might of the British Navy was constantly visible on the horizon.

When in January, 1815, the Americans won their greatest victory of the war on land at New Orleans the two-and-a-half-year-old war was over. The issues of impressment, the right of search, the maritime difficulties, and the restrictive Orders in Council were not settled; however, America had proven herself to be a formidable military power, even though the weakness of her young Navy was very obvious in the sea encounters with the British.

News of the Treaty of Ghent reached New York on February 11, 1815, and less than a month later Harmon Hendricks returned to his letter book for the first time in two years to transcribe the correspondence that signaled the reopening of the American copper trade with England. To his old accounts he extended congratulations "on the return of Peace," and without further reference to the war, asked for the release of the various goods he had purchased in the spring of 1812. In addition to these requests Cropper, Benson and Company, and John Freeman were sent supplementary, although small orders. Michael Oppenheim was brought up to date on the unending litigation with the Judahs, and to Peter Maze, Hendricks wrote what seemed to be a facetious confession:

I have been so long out of business that I shall not know how to bend my mind to profit & loss so I shall commence but very moderately for the present . . .

The resumption of trade was indeed modest; tin, brass, spelter, and the metals associated with copper were ordered in small quantities by Hendricks. It soon became evident, however, that caution tempered only a few of those associated with the copper trade on both sides of the Atlantic. English merchants were importing copper from Peru, which they were eager to resell to the United States, while withholding their own refined ores. The Americans, having experienced a wartime rise in copper of 60¢ per pound, had quietly hoarded what they could in an attempt to sell at higher prices. To this stock was added postwar purchases, but the speculators were no more successful with copper after the war than they had been with tin before the war. Peace, and the release of English exports, cut the price back to 27¢ a pound. As the price continued to drop Hendricks bought almost 250 tons, and was not anxious to increase this stock until he ascertained the potential postwar consumption of copper. Merchandise imported by the American merchants during 1815 consisted chiefly of iron wire, spelter, block tin, brass, and metal goods, which did not invite speculation. Thus, while Hendricks' warehouse was glutted with copper and choice metals, the speculators drove themselves out of the market.

Hendricks had every intention of returning to the carrying trade from which his original success was derived. Although the commerce across the Atlantic that had enriched so many of the New York merchants and shipowners had not lost its attractiveness it had been stripped of some of its profits. For Hendricks the trade across the seas continued to be an important source of raw and unfinished copper for the Soho mill and for the customers that he had faithfully supplied since the death of his father. As a manufacturer he was dependent upon imported metals, and as an importer he had one of the best outlets for his metals—the Soho Copper Works. Many of the New Yorkers returning to the carrying trade were not so favorably situated. The spread of American industry had upset the old source of wealth, and those who entered the race for riches had to find a new approach to business.

Of thirty-four New York merchants who had benefitted considerably from the carrying trade and other prewar business, Hendricks was among those whose personal estate in New York City was taxed on property valued at $60,000 or more. In 1815 the comparative wealth of these merchants ranged from a low of $5,000, recorded for John Jacob Astor, to a high of $200,000 for Robert Lenox. Two of the older copper merchants, Peter Curtenius and Peter Goelet, surviving contemporaries of Uriah Hendricks, were taxed on properties valued at $30,000 and $50,000 respectively. Richard Varick and Stephen Whitney exceeded the sum of $100,000, and Harmon Hendricks was on a par with the old tobacconist, Jacob Lorillard, and Archibald Gracie of the Chamber of Commerce, each of whom was taxed on the basis of the figure of $60,000. If the value of the New Jersey mill properties is added to his other property assets, then the worth of Hendricks' real estate was far in excess of either Varick's or Whitney's. In the next five years much of this wealth was toppled by the transformation of American trade and industry.

With the return of peace the bright prospects entertained for American manufactures dimmed. As soon as British ships could load and cross the Atlantic they crowded the American wharves and discharged their cargoes at prices that undersold the product of the New American factories. The superior quality of British goods made their sale easy. Imports from the Baltic cities also arrived in large quantities, and this mass of foreign goods competed so sharply with American-made goods that the growth of the young industries was temporarily curbed. A cry for protection was sounded. Heeding the protest Congress passed a tariff law in 1816 to help counter foreign competition. Russia goods, linens, and diaper sheeting were among the items subject to a double duty. The new tariff reassured the anxious friends of American industry. The Soho Copper Works expressed doubt as to whether a duty on copper would be levied, and to Paul Revere they stated, "there will be more honor in beating John Bull out of our market by the low price and superior quality of copper than by duties . . ." But the act of 1816 also placed a levy on brazier copper to assure the metal men that John Bull would be controlled in the American market.

The effect of the war was especially apparent in the shipyards. Once a thriving industry dependent on the carrying trade, shipbuilding declined when capital and labor were diverted into other manufactures. But shipbuilding was definitely on the eve of a new era. Steamboats were no longer viewed as one man's folly but were becoming an important innovation in maritime transportation. The imaginative shipwrights envisioned new styles for the vessels that plowed the ocean, or nosed through the inland waters, or sailed coastwise along the Atlantic seaboard. A merchant navy was about to be built. The United States Navy, in the realization that its hastily built, flimsy vessels had hardly been a match for those of the English, embarked on a program that would gradually increase its tonnage and strength from year to year.

The inducements offered by the government and the prospects of a merchant navy revitalized the shipyards, and once again Foreman Cheeseman, Christian Bergh, Henry Eckford and the Brown brothers turned to Hendricks for copper and cash. If there were any thoughts about Soho's serving only as a wartime expedient until such time as English copper was again available, these thoughts were soon dispelled. Shipbuilding renewed the demand for manufactured copper and promised much work for Soho. Importation of old and raw copper from England was, however, reduced, and the former status of copper imports was no longer the same. Sweden, Denmark, Russia, and the countries of South America competed for English and American markets. A quest for indigenous ore was begun by Hendricks, pursued by his successors, and continued by others until the United States became the largest producer of refined copper in the world. Hendricks' hopes for the new enterprise were high, and in 1815 the stocky, dark-haired forty-four-year-old New Yorker entered into the most interesting, if not the most profitable, phase of his career—that of a pioneer copper manufacturer.

II

The acceptance of steam as a primary power in water transportation in the New York area coincided with the opening of the Soho Copper Works. Shipbuilders, marine engineers, and foundrymen found that the conveniently located mill at Soho made it possible for them to obtain their supplies without loss of time. When it was discovered that copper could contain steam better than the square pinewood boilers that were used at first, copper became more desirable than ever before. Other practical uses of the metal, besides ship sheathing, were adapted by the engineers and found-rymen when they began the manufacture of steam engines and brass and iron machinery castings. The call for copper stripping, rivets, and boiler plate increased with the growth of steam navigation. The fruitful interrelationships among steam, engineering companies, and the Soho Copper Works developed simultaneously with the rise of American industry.

It was therefore not unusual for Robert L. Stevens and Robert McQueen, specialists in the building of engines, or John Youle, who made brass castings, and James P. Allaire, the marine engineer, to turn to Hendricks for copper. All of these men were associated with marine engineering, and at one time or another had produced metal castings for Fulton's steamboats. When Fulton's *Paragon*, the *Fire Fly*, and the *Demologus* were returned to the shipyards of New York for recoppering the metal was supplied by Soho. In 1816 Adam and Noah Brown, who did some of this work, purchased $10,000 in copper from Hendricks, and Henry Eckford's yard purchased copper in an amount that almost equaled that of the Browns.

In the years following the War of 1812 Hendricks became primarily involved in the three branches of the marine industry that were heavily dependent upon copper: steamboats and the engineering companies, sheathing for merchant vessels, and work for the United States Navy. Another branch in which he was involved, perhaps the most competitive, consisted in supplying the many and diverse metal needs of the general consumer. It was the latter with whom Hendricks had had the longest experience, and general sales had constituted the backbone of his trade until the United States Navy launched its shipbuilding program.

Taking full advantage of his experience in the carrying trade and his extensive list of customers Hendricks undertook to reach the nation with the products of Soho. Domes for public buildings and roofs for houses were now covered by copper; public water-works found copper essential, copper stills were more in vogue than ever before, and the demand for rivets and stripping had increased in proportion. The Maryland millers Tyson and Ellicott, the Custom House in Norfolk, Virginia, the New York State Prison, and grain merchants throughout the land placed their orders for sheathing, bottoms, and nails with Soho. In addition Hendricks extended his trade to the old South and reached out for the prosperous market of the new West.

The sharpest competition for the inland trade came from the merchants of Philadelphia. Of the three men who entered the copper trade after the War of 1812, and with whom Hendricks did business—George Harley, Robert Kid, and Nathan Trotter—Trotter warrants the most attention. An enterprising Quaker merchant, Trotter was soon to become the outstanding metal importer in the city of Philadelphia and subsequently a major competitor of Hendricks. His principles in the procurement of copper were identical to those of the New Yorker, and his methods of inland distribution were strikingly similar. Despite his sizable imports from England he appears to have carried on a large volume of business with Hendricks as soon as the Soho mill began rolling copper. Obviously it was necessary for Hendricks to maintain his agents in Philadelphia to handle orders for the mill and to meet the competition coming from men like Trotter.

The most important consumer of copper that Hendricks was eager to reach was the United States Navy. Here competition was limited to only a few, and bidding for naval work was a matter of public record. Early in 1816, in response to a newspaper advertisement, Hendricks offered 3,000 sheets of ship copper for delivery in Washington. To Commodore John Rodgers, president of the recently created Board of Naval Commissioners, the 12,000 bolt and spike rods and 125 cases of newly manufactured English copper were described according to naval specifications. Navy orders of 1816 differed from the sporadic purchases of 1804, or the hasty wartime purchases of 1812. An organized building program was under way that offered opportunities for copper manufacture which exceeded those provided by steam navigation. Under Rodgers' administration copper was incorporated as a standard naval requirement. However the Navy was not prepared to move hastily and Hendricks, aware of this, devoted his energies to the other phases of the marine industry and to the resurgent inland trade.

As a result of the rapid development of steam navigation and the wider application of copper in industry the Soho Copper Works surpassed Hendricks' finest expectations. Wages at Soho were unusually high, averaging $6,200 annually for approximately thirty mill workers and the two clerks employed at the Beaver Street warehouse. Compared to Revere's prewar payroll for twenty employees who altogether received $2,500 annually, Hendricks' payroll was about forty percent higher. In the absence of other records, further comparison is not possible.

For the hundreds of thousands of pounds of copper that were produced and sold annually, losses through refining were comparatively low. An average return of 37¼¢ per pound yielded about 10¢ profit per pound, but hardly a sufficient reward to the owners for their efforts. With the same unrelenting determination that led him to acquire huge quantities of metal from the intemperate speculators after the war, Hendricks decided to enlarge and improve the mill in anticipation of meeting the needs of the naval program.

During the summer of 1816 the owners of Soho erected a stone building, thirty by sixty feet, and walled the interior with firebrick burned in England. New machinery for turning bolts was installed and, to guarantee exactness in the rolling of sheets, Soho engaged the services of Oliver Evans of Philadelphia, the best American engineer available for the manufacture of new rollers. In his search for American ore Hendricks began to purchase or lease as many of the properties contiguous to the Soho estate in Belleville as the owners would permit. It was still believed that untapped deposits could be found in the once-productive areas mined by the Schuylers or worked by the Hornblowers. Almost no other information survives that could positively identify mining in the area other than the legal rights Hendricks obtained by deed for "all manner of mines, pits, and veins of Copper ore, lead . . . or other metal." American copper was surely found, for in the coppering accounts of Soho, ships coppered with American copper are distinguished from those that made use of old or foreign copper. Only one unresolved problem continued to disturb the men at the Soho Copper Works—their inability to reduce the quartz content in the refining process.

In an attempt to meet technological problems Hendricks wrote to John Freeman for advice that would enable his smeltermen to eliminate bubbling on the surface of copper and overcome the causes that led to the production of brittle sheets. Hendricks had established excellent rapport with his English correspondents. He was always ready to carry out personal favors or attend to business requests. English merchants were equally cooperative. But divulging the methods of smelting, refining, or rolling copper was not included in the relationship. If the Freeman company ever advised Harmon Hendricks of its metallurgical process there is no evidence of it in the Hendricks correspondence. There is a strong likelihood that the Bristol firm remained politely silent on the subject and the men at Soho struggled to devise their own process of metallurgy. Complaints received by Hendricks from his customers indicated that the surface of the sheets was blistered, or the copper was too brittle, or that it was rolled too soft. How these problems were overcome is speculative.

Despite the now more plentiful supply of copper England invited the world copper market to funnel its raw ore through her refineries. With the occasional exception of a direct shipment of Cuban or Venezuelan copper, South American copper reached the United States by way of England's West Indian back door. The establishment of smelting plants in Baltimore hardly upset this control, and the technological superiority of the English refineries enabled England to hold her dominant position in the copper market.

England was not so well situated in the control of other materials, however, and her exports declined considerably during this period of American industrial growth. Once again the American market attracted English attention. In the spring of 1816 tremendous quantities of English goods of every variety were poured into the United States. This time the quality of the merchandise was inferior to that dumped on the American market in 1815. The goods were accompanied by English agents, whose purpose was to falsify invoice declarations, misrepresent the value of the goods to the customs inspectors, and enter them at reduced prices, thereby lowering the duty rate. Once the goods passed the customs office they were sent directly to the auction rooms for immediate sale. This new flood of lower-priced English goods had made a mockery of the American tariff by circumventing it. By the time that the merchants, the manufacturers, and the retailers became aware of how their businesses were being undermined, they were almost destroyed. On July 1, 1816, less than three months after the hectic buying and selling of the imports had begun, Hendricks was compelled to write to Philip George:

Goods of all kinds are a drug here, I have all mine on hand waiting till the needy get rid of their intemperate importations. I shall not want any goods for the fall but should like to hear in Sep. or Oct.ʳ the Cash price of tin, spelter, Block tin . . .

And three days later he wrote to John Freeman:

. . . 2 years ago I had no idea that such times as we experience at present could come on us; such amounts sold 12 months ago to the Back country still remain unpaid and no new sales making save at auction. I have now the heaviest stock on hand I ever had without much prospect of reducing it to advantage or even at Cost. I have no expectation of wanting anything from Europe for a year to come.

The conditions that Hendricks reported to Freeman were also reflected in the mill's experience with the urban and country trade. From inland New York appeals came to Hendricks not to press for payment, for "the extreme scarcity of money" made payment for copper bought the previous year impossible; or he received reports that there was no money in the country; or that "not a single application has been made" in the West for copper. The same conditions were true in the shipyards, where enforced idleness was the result of the reduction of trade with the West Indies. Many of the shipwrights, for lack of other employment, went to New Brunswick to cut timber, and mechanics without other prospects were thrown in with the unemployed. Most of the merchants soon found themselves on the verge of ruin.

So severe and unexpected was the crisis that struck trade and

industry that if Hendricks had not earlier successfully secured
the profits of the carrying trade he, too, would have had cause
for alarm. His plentiful supply of copper was useless in an unresponsive market. Negotiations between Soho and the Navy
dragged on through the summer and fall of 1816, and it even
appeared as if the New Jersey mill would be forced to close for
want of work. Costly renovations, superior machinery, and the
investment in adjoining land consumed more funds than the business showed any immediate prospect of returning. The improvements begun when business was thriving were completed when
industry was on the verge of collapse and trade was forced to a
standstill. Further, in November of 1816, the Passaic River and
its tributary that led to the mill gave the appearance of an early
winter freeze, which would shut off all transportation to the mill
for four to five months.

If the Navy contract was approved after the rivers had frozen
its value would be lost. Transporting the hundreds of tons of coal
necessary to fire the furnaces and delivering the copper supplies
from Beaver Street to Belleville was as costly a risk as failure to
have the required materials in readiness. Hendricks determined
to take the risk. During the last three weeks of November the
men were set to work hauling 200 tons of coal and the necessary
quantities of copper through the streets of New York to the North
River, where it was placed on boats that carried it across the bays
and up the winding Passaic, then into the stream that led to
Belleville. No sooner was this completed than the Passaic and
Second Rivers began to freeze. The obstruction caused by the
early winter had been successfully avoided, and before the coal
was completely stacked in bins word came from John Rodgers of
the Navy Commission in the form of a counter-offer. Hendricks
was on edge, but his eagerness to obtain the order at a time when
there was no work and no sales did not prevent him from replying
coolly and politely that his estimate was based on cost without
benefit of any profit. ". . . that the differences between our offer
and the one offer'd by you bears no comparison with the danger
of ultimately shutting up our mills for want of work and encouragement to go on, we are therefore the more willing to accept
your offer . . ." Hendricks also reminded the commodore that

the government did not have to risk the dangers of the sea to obtain copper, or become involved in the English Orders in Council that still controlled the exportation of naval materials. In response to this frankly written letter the United States Navy approved the contract on December 30, 1816. All of the 264,751 pounds of bolts, spikes, and sheathing that went into the two seventy-fours and the one frigate, built the following year at the New York Navy Yard, were manufactured to meet government specifications by the Soho Copper Works.

Besides Hendricks, Paul Revere & Company and Levi Hollingsworth of Baltimore also competed successfully for naval work, but Hendricks appears to have received the largest number of contracts. The survival of Soho during the crisis that engulfed the United States in the postwar period was due to the contracts obtained from the United States Navy. Other phases of the metal trade languished. In Connecticut a large number of the tin traders failed as the demand for tin declined each day; iron wire dropped sharply in price, and copper sales continued to lag. The high price of grain, caused chiefly by the war excise tax, deterred distillers from purchasing new stills as long as they could repair old ones. The sheathing of ships was kept at a minimum, and the seaports were crowded with empty vessels. For the first time in many years the price of copper had fallen far below its cost in England. Hendricks' copper purchases became insignificant and his trade with England was reduced to the vital necessities that could not be obtained in the United States, such as specially fabricated wrought ladle molds used to scoop molten iron, firebrick for the mill, and coal from the English pits.

By the following year, when James Monroe became President ushering in the Era of Good Feelings, the American manufacturer had been fully confronted by the economic realities of the postwar period. Not only were the young industries forced to compete with a flood of inferior English merchandise, but with the effects of a worldwide depression in prices as well. According to George Dangerfield, a historian of that period, "this state of affairs was concealed by a land boom and a cotton boom." Dangerfield further states:

Farmers and shipowners, however, were also temporarily trapped by the reorganization of the postwar world. The ports of the British West Indies were once again closed to them, at a loss of at least six million dollars in agricultural exports, and a consequent unemployment of eighty thousand tons of shipping. British ships in the meantime moved freely from Great Britain to the United States, from the United States to the West Indies back to Great Britain; and this triangular trade, denied to the Americans by the closing of the West Indian ports, was immensely profitable.

For the Americans disaster was imminent. The decline in trade, commerce, and agriculture increased from month to month. The flush days that had kept the population of the Atlantic seaboard employed before the war now faded. The West and Southwest, which had enjoyed the boom, were now distressed by the crisis, and the sale of copper to the inland trade was reduced to a new low.

During 1817 these unpredictable conditions so weakened Hendricks' inland trade that the naval work assumed a higher status in the mind of the copper manufacturer. That same year the revolutions in South America, which affected the principal sources of raw copper and precious metals, propitiously affected the American copper trade. The metals of Mexico, Peru, and Venezuela became a medium of exchange in the hands of the revolutionists. The South American patriots came to Philadelphia to purchase supplies that would enable them to carry on the war of liberation from Spain. In exchange for the $250 million worth of goods they were eager to buy they offered copper and other metals. Hendricks and Isaacs lost no time in advising their English colleagues, Philip George, Pieschell and Schreiber, and Mather Parkes, who were commission merchants in the South American trade and who were in close contact with the patriots, to purchase whatever they could for direct shipment to New York.

Again Hendricks maintained his policy of purchasing copper in a depressed and unresponsive market, although he was unaware that at the end of the year changes in the shipbuilding industry were to take place. He stressed his preference for reddish Peruvian copper over the brassy-colored copper from Caracas, or the yellowish Mexican copper. Twenty years of importing South

American copper had taught him to recognize the superior quality of Peruvian ore. Hendricks replenished his dwindling copper stockpile at prices that seemed advantageous. In England the price of raw copper rose again, and its extensive use there provoked a belief that this was the harbinger of another war. Actually England was increasing her own naval strength when, at the same time, a slight upsurge in American shipbuilding became evident. With new naval contracts in hand, Hendricks required little additional encouragement to increase his stock. He estimated that the mill could produce 500 tons annually, and offered the English copper merchants £92 a ton for cake or pig copper. Other attempts were made by less discerning importers to sample the rose-colored Swedish copper, which was lower in price than English, and the dark, plum-colored copper from Russia, whose inferiority to the Swedish and Peruvian Hendricks easily recognized. Russian copper was limited to small sales in America. The selection of quality copper had become an art.

The constant need to improve upon the mechanical apparatus of the mill was a matter of equal importance. Buildings had to be repaired, refining and smelting procedures carefully watched, and rolling equipment observed for the slightest defect. American foundries had not yet acquired the English techniques of producing rollers of high quality. Revere turned to England for his first set; Haym M. Salomon failed in business because he was unable to correct difficulties in rolling; and Soho replaced its first rollers with a set manufactured by Oliver Evans. When these showed signs of wear Hendricks and Isaacs wrote to Thomas Holmes, Jr., of London for a new pair. Good rollers had to be made of wrought steel. "They must be turned down to 14 inches diameter and the whole surface true without as much as the size of a pins head on it, cast steel or common cast iron will not answer," Hendricks specified. New equipment had to be ordered months ahead, and although Hendricks would have preferred to have the new rollers shipped by the *James Monroe* or by one of the other fast-sailing packets, they were sent by the *Andrew Jackson* and lost when the ship was grounded at Barnegat Bay, off the New Jersey coast. The whole transaction then had to be repeated.

At the time the South American purchases were being negotiated, Hendricks wrote in the Soho ledger:

Total weight sold since the first sale made at Soho Mills for Philadelphia 469,738 lbs. for $175,682.75, total average 37¼ cts per lb.

Profit for the first three years hardly amounted to $30,000 which, when divided equally between Hendricks and Isaacs, was not quite $5,000 per year. For Hendricks this return was slight in comparison with his returns from banking, real estate, and stock investments. But he was determined to persist in this business although it demanded time, caution, and constant reinvestment. At the beginning of the crisis of 1816 his resources were such that he was already counted as one of the wealthiest New York merchants. If he had decided at that time to surrender all interest in the manufacture of copper his financial status would have remained undisturbed. Isaacs, if necessary, could have taken over the operation of the mill, which he was managing efficiently. But copper had become a love with Hendricks, and it was already part of his family history. Since the end of the war this attitude was reflected more and more in his correspondence with English merchants, or with the Navy commissioners in such statements as: "the former house I dealt with 20 years and my father before me 30 years"; or, "I have sold copper 25 years"; or, "my ancestors & myself for more than 70 years have turned our trade exclusively to copper." The old generation of New York copper men, the Goelets and Curteniuses, had passed on, and only one newcomer, Peter Harmony, whose name was later to be associated with transportation history, entered the New York copper field. The Hendricks copper firm was at this time the oldest in the United States.

The encouraging upsurge in shipbuilding that began toward the end of 1817 and continued through 1818 was important but short-lived. The Navy had undertaken to build a series of battleships for which Hendricks supplied 100 tons of copper. Partly because of the improvement of the packets, the pioneer attempts to adapt steam for the use of ocean-bound vessels, and the competitive state of foreign trade, there was a revival in building. The

Coffee House Slip and New York Coffee House, 1856.

new packets were larger, swifter, and designed to carry both passengers and freight. Once a month these fast boats, operated by an American line, left New York for Liverpool, and left Liverpool for New York as regularly for the return trip, instituting the custom of scheduled sailing, and rapidly expediting trade and mails across the Atlantic. The advent of the monthly packet brought to an end the old system of spring and fall shipping.

Some of the better-known packets made use of Soho products. Copper went to Adam and Noah Brown for the *James Monroe;* to Sidney Wright, whose name is intimately associated with the building of packets, for the *Albion;* and to other shipbuilders for the *Courier* and the *Pacific.* Hendricks took pride in advising his Liverpool correspondents to use the packet lines and, whenever possible, he imported goods on these ships, or on the *Mars,* the *Phocion,* the *Euphrates,* and the ill-fated *Andrew Jackson.*

In the summer of 1818 Fickett and Crockett's East River shipyard, where many fine coasting packets were built, undertook the construction of an entirely different type of vessel. It was an epoch-making venture that resulted in the building of the first steam-propelled vessel to cross the Atlantic Ocean. The rods as well as the boiler sheets, and possibly the sheathing, were made by Soho for the 90-horsepower steamship. Today, its name, the *Savannah*, has been adopted for the world's first nuclear-powered cargo-passenger ship. In addition to the *Savannah* one other ship, the *Robert Fulton*, important to the history of steam navigation, was built by Henry Eckford with Soho products while the *Savannah* was on the seas.

This short-lived boom in shipbuilding had no relation to the growing crisis that gnawed at the American economy. Hard times were widespread. For more than two years the banks had been gradually contributing to the depressed conditions by flooding the country with worthless paper money, increasing the distress, which by the spring of 1819 turned into a national panic. Hendricks' banking interests placed him in a knowledgeable position about paper money speculations. The Hartford Bank, with which he was closely associated, had secured its specie and had sound currency. Even such unquestioned collateral was not enough to ease the problem of paying bills. Philadelphia merchants refused to accept payment from Hendricks if a bank was personally unknown to them. Notes issued by a bank fifty miles distant were looked upon with suspicion, and drafts were considered worthless until they were deposited and cleared for payment. More and more the banks were held responsible for the Panic of 1819. Bills

The *Savannah*, first ocean steamer.

abroad were more difficult to pay, and Hendricks was compelled to explain this predicament to John Freeman in Bristol:

For the first time in my business, I have now found a difficulty in selecting bills in England. Those houses which are estimated substantial refuse to draw for fear their correspondents may have failed in England. Those I can procure I cannot touch for want of confidence in the Drawers—I have promised to remit to you, have therefore been compelled to remit you the enclosed United States 7 pr ct stock certificate No. 716 dated yesterday in my name for 7376.79 with my powere attorney to you, to sell . . .

The financial system was throttled, unemployment was unusually high, and with bankruptcies a common and daily occurrence, business was at a standstill. To all of his correspondents Hendricks gave a vivid description of the deteriorating state of trade. Sales for cash could not be made, and no careful merchant would hazard sales for credit. For the first time since the War of Independence, the American workingman became dependent upon such compassion as could be found in the newly organized soup kitchens, fuel and sewing societies. He became a needy recipient of food, clothing, and shelter from the charities that arose almost overnight through the efforts of the bountiful of New York City. Hendricks' personal friends turned to him for assistance, and at no time in the past did he recall seeing such abject distress in New York.

The shipowners who survived the curtailment of trade in 1816, who patiently withstood the languid months of 1817, who looked for a revival of trade across the seas in 1818, finally succumbed to the banking crisis of 1819. Owners of real estate converted property into cash. Merchants offered their vessels for sale rather than watch them rot on the wharves, and holders of negotiable stocks and bonds were forced to relinquish them. This anxious market invited Hendricks to purchase large quantities of notes and stock, whose soundness the most cautious judge could not doubt. The same careful selective technique that he had applied to the purchase of the hoarded tin in 1811 and the glutted copper piles after the war he now applied to as much real estate as he could afford. By the time that summer arrived and the huge Navy orders were completed failures in business reached a new high and the merchants could speak of nothing else but empty ships. "Ninety out of 100 ship holders and shippers of cotton & corn must fail from the very bad returns from Europe," wrote Hendricks in appraising the state of the shipping trade. Richard Varick lost $30,000; Robert Lenox and John Hone lost $40,000 each; Archibald Gracie was reduced to $20,000, and many of the once wealthy merchants were totally impoverished. Whitney was among those in the ascendant. John Jacob Astor still hovered in the background, and Hendricks showed an increase of $20,000 in real property.

Yet no panic or economic crisis was able to deter young and energetic newcomers from entering an arena from which older and more experienced men had just retreated. Another Gomez, unrelated to Hendricks' French brother-in-law, but one of the extensive New York clan, entered the West Indian trade and soon after found himself in the employ of Hendricks. The packets from Liverpool that brought English goods also brought the enterprising, flamboyant Tobiases, who became associated with the Hendrickses shortly after they set up shop in New York. In Philadelphia, David Seixas was busy scratching an engraving on a piece of polished copper that would display his invention of a language for the deaf and dumb. He permanently withdrew from the copper trade to found the Pennsylvania Institution for the Deaf and Dumb. When he did so, a member of the prominent Moss family was ready to take his place.

With no ships to copper and with the government work completed, Hendricks, on learning that the Navy was about to build three steam frigates, once more addressed Commodore Rodgers:

Since we finished the last parcel of our contract for the United States, we have not had work enough to keep our men employed & to pay for opening our mill. We humbly suggest to the Navy Board the favorable time to make a contract with us. We should take 5 or 600,000 ls. copper lower than the last rather than discharge our men at present and shut up our mill.

Instead of discarding its old copper the Navy sent it in to Soho, where it was exchanged for manufactured copper. Hendricks' bid to manufacture copper castings and the boilers for the frigates fortunately was successful. As the bitter winter of

1819 brought with it more distress than relief, he gradually reduced his copper purchases from five to two hundred tons annually, but kept the mill in operation. The appeal to the Navy for work implied that Soho needed work, but it was not in distress. Navy contracts for that year had amounted to $34,000; miscellaneous sales for a fourteen-month period from the end of 1818 through 1819, though low, kept the mill workers regularly employed, for the $9,000 payroll for the same period had exceeded the average annual payroll of the preceding years. The net profits were still slight, yet Hendricks and Isaacs were content in the fact that the business was kept alive. Undeterred, Hendricks continued to produce sheet copper and bolts. In the course of manufacture, caulk or pickle dust—pure copper by either name— became a by-product. Chemists used it for blue vitriol and coloring, and though the Reveres were large purchasers of the product Hendricks attempted to export it to England when a satisfactory price could be obtained.

III

If the Hendricks' ledgers are an accurate record of the times American shipbuilding resumed its former activity during the summer of 1820, when Navy contracts were again awarded and the foreign trade slowly reawakened. An increase in trade was enough to prod the shipyards back to life. At the yards of the oldtimers, Christian Bergh, Foreman Cheeseman, and Henry Eckford, the ring of the coppersmith's hammer could be heard again. Noah Brown had just finished building the *China* for Samuel Hicks; the steamboat *Paragon* was freshened and repaired; Fickett and Crockett, the Wrights, and the Webbs all showed signs of renewed activity as the packets and merchant vessels were refitted for use. The sea lanes between New York and Bristol, Liverpool, and London again were churned by the vessels of the American merchants. Writing John Freeman about this progress in shipping Hendricks observed, "Our American ships have wings, while your British Packets are generally dull sailers."

Recovery was slow, and words of assurance could be heard in the counting offices of New York. Columns of maritime news, ship arrivals, and departures filled the press and gave substance to the mercantile gossip. Congress, still reacting apathetically, considered a number of laws that would bring relief to the large numbers of bankrupts of 1819, and studied the possibility of a new tariff. While one of the new measures, a general insolvency act, was being debated in the early months of 1820 Hendricks reviewed the status of those heavily indebted to him. He informed John Freeman, from whom he had obtained power of attorney to collect the Bristol company's American debts, that if the law was enacted he would be unable to make collection. By the time Freeman received Hendricks' letter another was on its way, stating that Congress had not passed the bankruptcy act. Debtors became as elusive as hard cash; one died, another disappeared, and still others justifiably pleaded mercy. There were very few suits, and few of these recovered any cash. In Hendricks vs. Robinson, a complicated case that also involved the Judahs, large parcels of timberland in Canandaigua, New York, were awarded by the court to Harmon Hendricks.

Congressional considerations for a revised tariff included a heavy rise on copper. Disturbed by the possibility of the new law Hendricks at once cancelled all copper shipments scheduled to leave England after May 14, 1820, and awaited the result of the new legislation. Meanwhile he prepared a petition to be sent to Congress that would clearly voice his protest against the existing law and encouraged necessary revisions. For almost fifteen years the tariff on copper had been a source of unresolved annoyance to him. There was the old, unsettled matter of 1806 that awaited action by the federal court. The terms under which the custom officers at New York City collected duties were still judged as ambiguous. In 1809 a partial refund of duties was made to Hendricks by the Collector of the District of the City of New York on the importation of flat bottoms and bolt rods, but the whole matter remained unsettled. Another petition by Hendricks, urging a refund of duties on imported copper, was addressed to the Senate and House of Representatives in November, 1811, and it was referred in turn to the Committee on

Commerce and Manufactures. Again the issue was indefinitely suspended and the war conveniently pushed it out of sight. New laws, revised tariffs, a disrupted world market and the desire to completely reorganize the domestic copper industry on the part of individuals in the trade brought about the petition of 1820.

Hendricks' new petition differed in need from his former one, since his position was now somewhat changed. The prewar petitions had been directed to the welfare of a shipping merchant who was active in small manufacturing, but now Hendricks' petition concerned itself primarily with the safeguarding of manufacture. Hendricks, writing for Soho as a smelter, a refiner, and a manufacturer of copper of every description, briefly told the story of the founding of Soho in 1813, its success in producing "copper of the first quality," and the struggle to keep the mill in operation:

. . . an abandonment at this time would be attended with a heavy sacrifice, their buildings and machinery being useless for any other purpose than for the refining and smelting of copper.

Your memorialists further represent that they are competent in conjunction with the other manufactories of copper in the United States to supply the consumption of [the] United States; that they have manufactured many tons for the use of the *Navy* of the *United States* and that the aforesaid calls, for the use of the navy have been the means of keeping their workmen together, and their establishment from being shut up: but now that the Navy of the United States is supplied with the copper for the increase of the Navy, they will have to look for the other means of employment, and for protecting duties at the hand of Government against foreign importations of copper.

The key to Hendricks' request for increased protection of American copper lay in his assertion of the improper declaration of the uses of copper by importers. Sheathing copper remained exempt from duty because of its naval use. But, according to Hendricks, one-third of the copper sheathing imported into New York State was diverted to other uses, to the detriment of both American revenue and the copper manufacturers. Contrary to all expectations Congress did not impose the additional duty on sheathing nor the hoped-for 10 percent on other goods. The bill was lost by one vote. While this matter was under discussion

Henry Meigs, the New York congressman, was authorized to inform Hendricks that his nine-year-old claim for a return of duties paid on copper bolts had been adjusted by special congressional action.

During the debates in Congress Robert Swartwout, the New York Navy agent, applied for a quantity of pig copper. Swartwout incorrectly assumed that pig copper, when melted, could be used directly for ship work or the manufacture of cannon. Hendricks carefully explained to him the smelting and refining process and solicited the order. For the balance of the year whatever work he received from the Navy was insignificant. Two more years passed before the naval program entered into its next phase, but meanwhile Soho sought the work of the shipowners and the inland steamship companies.

Although Hendricks was a shareholder in the companies formed to promote steam navigation he did not become involved in the Fulton and Livingston monopoly to control steam navigation on the interjacent waters of New York. When Governor Aaron Ogden of New Jersey turned steamship promoter and challenged the monopoly, Hendricks, from whom Ogden bought his copper, wisely remained in the background. Hendricks was as dependent on Ogden's steamer *Sea Horse* for transportation between Elizabethtown Point and New York City as both Ogden and Fulton and Livingston were dependent on Soho for copper. Before the monopoly of the New York waters was broken by the Supreme Court of the United States in 1824—in John Marshall's important decision in Gibbon vs. Ogden—rivalry among the various companies only tended to further advance the interest in steam navigation. To keep pace with the astonishing developments in transportation new shipyards were opened near Corlears Hook on the East River, and two new accounts, those of Isaac Webb and Isaac Wright, were added to the ledger of Harmon Hendricks.

As the older vessels came in to the reactivated shipyards for repair Soho received sufficient work to remove all fear that the mill would close down. The older shipyards showed unquestioning signs of recovery; their new ships lay on their stocks, ready for launching. At Corlears Hook, the "Salt Side of New York," large piles of freshly cut white oak, locust, and cedar trees could be seen. Ship-stores were loading, the sailors' rendezvous along Water and Cherry Streets hummed like a hive, and young tars busied themselves filling their chests with seafaring equipment. The new designs of the ship architects were visible in the longer, leaner craft, and the imaginative builders liberated their ships from seagoing impediments. The new models were also distinguished for their speed. New yards were built, and launchings were a regular sight on the rivers of New York. These yards now rivaled the best of New England, once the cradle of American shipbuilding.

The eight major shipyards of New York, the coppersmith, and the commission merchant who dealt in metals, in a port destined to become the largest in the United States, had grown dependent on Hendricks and the Soho Copper Works. The convenient location of the New Jersey mill, its ability to manufacture promptly, and its facilities for transportation placed it in a unique position. Hendricks' willingness to extend credit won friends and customers for him. The experiences of the preceding years were fresh in the memories of those whom Hendricks had not turned away. His reputation extended the length of the Mississippi and along its tributaries, and he was known in the shipyards from New England to the South.

The year 1821 opened with modest requests for coppering and repairing the steamboats *Mexicana*, the *Richmond*, and the *Chancellor Livingston*. Allaire's large marine engine works negotiated contracts with Soho, and before the year ended every yard in the area was making huge demands for Soho copper. The forges and foundries were working full blast. In March the Gray Brothers of Louisville, Kentucky, were concerned about copper for their two steamers, the *Fayette* and *Shelby*, that plied regularly from New Orleans to Louisville, and they were also concerned about meeting the Kentucky demand for copper. In May Sidney Wright began building a new ship for Francis Thompson and required thousands of dollars in copper. In July Leroy, Bayard & Company placed their order for 1,200 sheets of American refined copper, which they specified had to be equal in all respects to the quality furnished the Navy. During the same month the Quaker Twains of New Bedford ordered 1,800 sheets of copper, warranted to last for no less than seven years, for the two boats that they were building. In August Noah Brown called for copper for a United States lightship. In September James De Wolf, the prominent New York merchant, requested the manufacture of sufficient copper for a 300-ton vessel building in Bristol, Rhode Island. In October it was Isaac Wright who asked for a similar quantity of ship copper, and in November Robert McQueen ordered copper for the boilers of the steamboat *Connecticut*. Hendricks himself never shook the influence of the carrying trade, and he entered into a joint venture in building the *Empress*, a coastwise vessel constructed by Foreman Cheeseman. The *Empress* was outfitted and supplied with a supercargo to be sold for rum and molasses in one of the islands or bartered for a quantity of candles.

The year 1822 repeated the good fortune of 1821, and for the remainder of the decade the maritime industry boomed. Older ships, such as those of the North River Steam Boat Company, made frequent demands for copper; the steamboats *Hoboken* and *New York* requested American copper; the *Robert Fulton* came in for coppering and new boiler sheets; Fickett and Crockett and Charles Brownne placed their customarily small but regular orders; Samuel Hicks, the South Street commission merchant who had had his nose tweaked by Hendricks fourteen years before, ordered a mass of copper for his ships, the *Galaxy* and the *Baltic*, and Christian Bergh placed an order that would copper the 260-ton *Don Quixote*.

The advertisement of a coppered ship wrote Herman Melville in *Redburn*, "suggested volumes of thought" to a young sailor. "*Coppered* and *copper*—fastened! That fairly smelt of the salt water! How different such vessels must be from the wooden, one-masted, green-and-white painted sloops, that glided up and down the river before our house on the bank." Wooden ships and iron ships involved the imagination of the novelist, of the literary critic and historian, but only Melville was infatuated by the romantic image of a coppered ship when it still plowed the sea.

Ships of every description demanded copper. It was the accepted mode for the protection of ship bottoms and essential for steamships. The sale of a new ship or an old one was as dependent on the condition of its coppering as on any other feature. Attempts to rival copper with zinc sheathing failed, and when the Navy resumed its building program in 1822 orders for copper flowed in regularly for the next four years. The Navy yards at Portsmouth, New Hampshire, Boston, Brooklyn, Philadelphia, and Gosport, Virginia, were enough to keep Soho busy. The schooner *Shark*, the brig *Enterprize*, the *Nautilus*, and the *Constitution* consumed large quantities of specially fabricated nails, rods, and sheets. James Kirk Paulding, the novelist turned Navy commissioner, negotiated the contracts between Washington and Soho. Commodore David Porter, a hero of the War of 1812, approved contracts for hammer-hardened bolt rods for delivery to Commodore Isaac Hull of *Guerrière* fame, who now commanded the Navy yards at Boston and Portsmouth. There was no longer any question about the growth of the Navy of the United States.

The crisis that had earlier squeezed the copper importers out of business and forced some of the young metal manufacturers into bankruptcy had left the field open to Hendricks. In the spring of 1822 he wrote, "there is not one importer of 3 years standing in the copper trade at this time," and, my "policy has been to worry other persons out of the trade." When economic conditions were good and the competition increased, Hendricks' experience, his wide variety of assorted coppers, his financial resources, and his European connections enabled him to maintain control of the copper market. He explained these circumstances to Mather Parkes and Company on one of those few occasions when he wrote of himself. In that same letter he expressed his impatience with the merchants who bought copper for their ships but showed little knowledge of what they were buying. He may even have recalled the Navy agent who was unaware of the need to refine copper before it could be put to use:

A merchant calls on me wants copper for his ship 10 to 1 for the 1st time without knowing copper from Brass. Have you first quality sheathing? Yes. Do you wish copper such as A B C D & E have on their bottoms, running 4, 5, 6, years on the copper. Whats your price? 26½ cts pr lb. It's ½ cnt above Mr. W. price. His copper probably is such as the ships F G H & I have had on their bottoms 2 to 2½ years, and was supplied with my copper last fall. I have the same worth of it—You may have it at 24 cts but I cannot warrant it to last 2 years, thus you see I have the power to use up any copper I have, and have had the power to ware out every new adventurer in the Business . . .

The varieties of manufactured copper were as important as the proper selection of the mixed ores from which they were refined. Losses in refining were as perplexing as they were constant. Expert smeltermen were not common, and it is not even certain whether the only smelterman in the area worked for Soho. The secrets of metallurgy were shrouded in the same secrecy as the alchemist's art. Hendricks and Isaacs became their own metallurgists, Hendricks studying the refining losses in detail, and seeking information from the experts in Bristol.

Ship copper received its own designations: there was English, American, old, and Soho copper, the latter possibly coming from the old Schuyler mine, which was worked only at intervals. Each was described according to its malleable quality or its brittleness, and these characteristics determined the length of warranty given to shipowners, which ran from two to seven years. English copper, refined by Freeman in Bristol, became noted for its strength, and had the reputation of holding to ships for seven years. Soho copper held for five to six years. Old copper, and copper refined from scraps and clippings mixed with the Soho product, was warranted to hold from two to four years. Refunds were guaranteed by Hendricks for all ship copper that did not last the warranted length of time, provided the ship had not been grounded and could be brought into the port of New York for inspection. Copper warrants were introduced in the mid-1820's when the copper market was happily supplied, when prices in England were lower than those in the United States, and regular shipping made copper available from the centers of its production.

English competition for the American market did not diminish. New applications were sent to Hendricks regularly. Notable was that of Newton, Lyon & Company who acquired the Flintstone mill of the old Marquis Anglesey mine, and who at one time had envisioned a world copper monopoly. Now they were anxious for

Hendricks' patronage. Despite the reputation of this old firm Hendricks proceeded to deal with them as if they were beginners, introducing them to his methods of business. These were: the seller required cash in advance with an order in United States bank stock, specific shipping instructions that were not to be altered, and the manner of insurance. Above all, he always included a reminder that his reputation in the United States could be attested to by any merchant in New York or by any ship captain who entered an English port.

At this time the only interruption of trade, the seasonal recurrence of the yellow fever, was on the American side of the Atlantic. Tariffs now favored heavy goods, commercial shipbuilding outpaced the Navy program, and the transatlantic trade no longer hungered for cargo from either side of the ocean. European demands for the products of the American fields—cotton, grains, and tobacco—were met by the swift-sailing American vessels, which returned with articles from the European factories. But the lighter goods from Europe left ample space for the emigrant who sought a corner in the ships sailing for America. The English factory system displaced the workingman, whose life was temporarily disrupted by the introduction of machinery. He now looked to America. Thousands of emigrants from the British Isles and western Europe struggled to get passage space on the ships that sailed to the United States. As a result the shipbuilding industry continued to thrive.

For the third time in ten years the facilities at Soho had to be enlarged. The adjoining cotton mill was purchased and incorporated into the Soho tract where the lower, or "Soho" mill was located. The original Schuyler mill was enlarged, but the pressing demand for manufactured copper quickly outgrew the new facilities. Another mill was erected in 1824, and it became known as the Upper Mill. New equipment was installed at the cost of $20,000. By tapping the waters of the Second River an additional sixty-horsepower system was developed. Only the losses in refining copper still worried the men at Soho. To expedite the delivery of copper to and from the mill, two schooners sailed regularly between the rivers of New York and the Passaic, carrying copper from the Beaver Street warehouse to the mill at Belleville.

Earliest known photograph of the Soho Copper Works.

No sooner were the new buildings completed than thought was given to plans for a new homestead on the estate. The outmoded living quarters had already been remodeled for the convenience of Isaacs, who traveled from New York to Belleville, and for Aaron L. Gomez, who was a permanent resident. In 1826 the building of a spacious home was begun. Fronting the Passaic River to the east, and overlooking the copper works to the west, the homestead was erected on the site of what later became the Forest Hills section of Newark, New Jersey.

Expansion and improvement of the mill area became a familiar sight to the townspeople and the mill workers. Soho wages sky-rocketed from a payroll of more than $15,000 for the period between March, 1822, and August, 1824, to double that for the period from August, 1824, to January, 1827. After the Upper Mill was placed in operation copper production reached 350 tons per year. Belleville had grown into a thriving community, dependent upon a major industry for its livelihood. Workers deserted the New Jersey countryside for the advantages of regular employ-ment at the copper works; mechanics were attracted from New

Hendricks summer residence at Soho, New Jersey, built in 1826.

York by the higher pay at the mill; and the farmers of Essex County searched their land for ore that could be mined and then refined at Soho.

American copper mills meanwhile had increased in number and entered the race for naval work. The Gunpowder Copper Works of Baltimore, the Crocker Brothers of Taunton, and the Swifts of New Bedford, each of whom bought copper from Hendricks, vied with Soho. The Reveres were no longer the only New England copper mill. Of all the mills, the largest single naval contract had been received by the Soho Copper Works. The proximity of the mill to the busiest network of rivers and canals in the East and to the best natural port on the Atlantic coast, enabled it to benefit from the advantages offered by a vast shipping trade. The opening of the Erie Canal further enhanced this advantage. Soho's reputation brought ships like the *Savannah* from great distances to be resheathed with New Jersey copper, and the satisfactory fulfillment of the naval work encouraged the United States Armory at Springfield to purchase tons of copper and zinc for the production of its ordnance.

The quality of American copper was still debatable. Yankee shipbuilders of New Bedford thought it too brittle; the downeasters stubbornly preferred English to American copper and the Nantucket yards gave preference to the blister-free product of the English rolling mill. Furthermore, William T. Grinnell, in the name of the New England shipbuilders, wrote: "I wish you to know how our *Yankees* talk about your copper, notwithstanding the statement about the Canton's copper of your make. Our good people will have their own way." It made little difference if the copper rolled by Revere was refined by Hendricks; prejudice against American copper persisted, but the majority of New England orders still were sent to Soho. In Baltimore the prominent miller Isaac McKim turned copper manufacturer when he succeeded Levi Hollingsworth, and he attempted to capture the whole market from Maine to Mississippi. Harrison and Sterret reported that McKim's venture was "actuated by Pride and Ambition, more than the prospect of profit. He would be the greatest manufacturer in the United States! He is, you know, very wealthy, but he is not adding to the heap by his copper operation." The Baltimore merchants remained faithful to Hendricks long after McKim vanished from the field of copper.

Soho was now so firmly established as a result of its modern mill, rapid production, and competitive prices that it easily surpassed all others. When Soho reached this peak in its brief history it became the first American copper mill to undergo reorganization, and then it exceeded even the highest expectations of its founder.

IV

The copper world that had assumed a major importance in the career of Harmon Hendricks did not distract him from his private life. Communal responsibilities, affairs at Congregation Shearith Israel, and particularly his growing family all received a large share of his attention. The younger boys were busy with their tutors and studies, and the older ones, Uriah and Henry, were fascinated by tales of the copper trade. The girls played heartily on the mahogany piano that their father had long ago purchased from Michael Oppenheim, and when company filled the spacious Hendricks home it was considered proper entertainment for the girls to display their dancing skills. And so the Greenwich Street household kept Frances Hendricks alert and her servants busy. Her brother, Solomon Isaacs, who gave all appearances of remaining a bachelor, lived with them when he was away from the mill. Visitors were frequent, and the warmhearted, talkative Tobiases spent many evenings at the Hendrickses gossiping about the Jews of Liverpool. The family's life was marred only by the death of four-year-old Fanny in 1817, which occurred within months of the birth of Hannah Hendricks.

The 100,000 inhabitants of New York City included only a small increase in the Jewish population, which in 1817 was approximated at 400. It was enough, however, to swell the ranks of Shearith Israel, still the only synagogue in the rapidly growing city. The Mill Street edifice, in which three generations of New York Jews had worshiped in the course of almost a century, was showing its age. Moreover many of its members no longer lived in the immediate vicinity and had to walk a longer distance to the synagogue. Most of the members obeyed the prohibition against traveling on the Sabbath and Hendricks, who lived two miles north of Mill Street, was among them. The yellow fever epidemic, which had prompted the first suburban movement out of the tip of Manhattan as a temporary refuge for those who could afford it, had now, as a result of its annual visitation to the city, established suburban Greenwich Village as an area of permanent residence. Indirectly it may have motivated the proposal to establish a branch of the congregation in the Village. The prosperity that characterized the supporters of this innovation, Hendricks included, surely contributed to the idea. Yet despite the commercial development that enveloped the Mill Street section it was decided in 1817 to rebuild and enlarge the synagogue on its original site.

The new plan was supported by a $250 cash gift from Hendricks. This was supplemented by a loan of $1,000, free of interest, for a fourteen-year term. These benefices were bestowed with the same lack of ostentation that characterized all of Hendricks' business transactions. When the synagogue was rebuilt and ready for dedication, invitations to the ceremony were sent to the highest American dignitaries, local and state officials, including the mayor of New York, the governor of the state, and the Vice-President of the United States. Leading non-Jewish laymen and a number of Christian clergymen who were present witnessed the traditional Jewish ceremony of knocking at the door for admission, and its opening by Harmon Hendricks. It was the first time in eighty-nine years that this ceremony had been seen in New York. In the procession that followed within the synagogue Hendricks participated in the seven circuits around the reader's desk, carrying one of the nine Scrolls of the Law. Upon the completion of this ceremony the congregation settled down to listen to an hour-long discourse by the colorful Tammany orator Major Mordecai Manuel Noah.

Heading the committee on arrangements for the dedication was Moses Levy Maduro Peixotto, a learned Curaçao merchant who had settled in New York in 1807. Hendricks and Peixotto became warm friends, shared their knowledge of the enterprising West Indian Jews, and discussed the affairs of interest to a Jewish community on the eve of change. Peixotto occasionally assisted Gershom Mendez Seixas with his rabbinical duties, and upon the death of the latter in 1816, succeeded him in the ministry. In addition he found time to supplement the Hebrew education of the Hendricks' children.

Their basic Jewish education was received at Shearith Israel, where the children attended faithfully, but it was their father who imbued them with Jewish precepts. If they did not acquire a scholarly knowledge of Hebrew, they possessed considerably more than a passing knowledge. The daughters were not cut off from this area of learning, although it was not yet customary to introduce girls to the study of Hebrew. Selina, born when Harmon was fifty-two, developed such a fondness for the language that one of her pastimes was transcribing Hebrew liturgical verse into her commonplace book.

Rabbi Moshe, as Hendricks fondly referred to Peixotto, was even induced to trade in copper from time to time and became a signatory to a number of the naval contracts. During the Rabbi's twenty-one year residence in New York the two men became staunch friends, and at Peixotto's death in 1828 Hendricks paid the fine required by the city for burial in the old Chatham Street cemetery.

In the decade that followed the War of 1812 a new stratum of Jewish society, largely descended from early Spanish and Portuguese settlers, began to emerge in New York. Wealthier than the average Jew, American-born, active in Sephardic Shearith Israel, and closely identified with English and Dutch Jewry, the new elite was small in number. Among this older segment of Spanish and Portuguese Jews some were not quite so fortunate. Hard times had fallen upon some of the once-wealthy Gomezes, the Lopezes, and the Isaacs families, and pride prevented them from turning to either the synagogue or one of the charitable societies that could have aided them. Instead they turned to Harmon Hendricks, whose purse was as dependable as the promise of his confidence. His response to personal appeals was always without fanfare or prejudice, and the details of his charity were never recorded for posterity. An occasional docket, or a note on the margin of an imploring letter are all that survive.

His attitude to Isaac Gomez, Jr., the poet, anthologist, and schoolmaster, was somewhat different. In 1813 Gomez had proposed the construction of a house to accommodate the Jewish poor of the city, and a few years later he was in such desperate personal need that he was forced to appeal to Hendricks. Hendricks was fully aware that extending a business loan to Gomez to give him a fresh start would not solve his friend's problems. After arranging for Gomez to teach in the congregation's school, he added the following note:

A friend should give his opinion freely, therefore, let me say that I consider your abilities equal to the task; but not equal to mercantile affairs that require ways and means independent of a large *Capital.*

This degree of family aid expressed a basic attitude of religious conduct. It was also an attempt to uphold and maintain a social structure that was no longer characterized by wealth, by those who remained wealthy. This was especially characteristic of the Jews of Spanish and Portuguese origin.

The decline in trade that had begun in 1816 and culminated in the Panic of 1819 resulted in broader communal responsibility. The economic crisis created such poverty that the Jewish charity chest was soon emptied, and the lot of the Jewish poor was not unlike that of the other poor of the city. But the organization of Jewish charity was considerably different from any other group's. Steps were taken to prevent public beggary by Jews, and organized efforts were made to care for the poor and the immigrants. There was no need for the Jew to depend upon the public soup kitchen for food, and only in uncommon instances were Jews to be found in the city's poorhouse. Hendricks revealed a deep interest in social welfare, not merely as a benefactor, or as the wealthiest Jew in New York City, but as an individual expressing a mode of life and the fulfillment of a religious ideal.

It is a curious fact of New York history that the yellow fever of 1798, which prompted the first suburban movement among Jews, coincided with the organization of the first separate Jewish charity in the city, the *kalfe zedakah matan beseter.* This charitable society undertook to supply those needs that could not be met by Shearith Israel when the raging yellow fever disrupted the community. Perhaps the death of Harmon's father during the fever of 1798 and the difficulty he encountered at the time of the burial prompted him to become a prime mover of this group. In any case, he was a prominent member of the society as long as it existed. The decline of the *kalfe*, whose funds were distributed in secret, can be connected with the decline of the fever in New York City.

One of Hendricks' benevolent endeavors at this time was his attempt to help regulate the price of *kasher* meat by subsidizing the slaughterer's fee. The object was to make lamb more readily available to those who could not pay the slaughterer's charge. Though this personal attempt to reduce the cost of *kasher* meat for the poor was not successful, it revealed the social vision of a man who might easily have chosen to remain aloof from such problems.

Beyond the immediate area of Jewish philanthropy Harmon Hendricks had not lost interest in the Society for the Reformation of Delinquency, of which he had been a member since 1806. As far as the record indicates he contributed to it annually during his lifetime. This liberality was not restricted to New York City. When David G. Seixas organized his second school for teaching the deaf and dumb in Philadelphia the largest single gift came from Harmon Hendricks, who not only became a life member, but donated $100 in addition.

The care with which Hendricks hid his extensive gifts from the public eye, frequently even giving them anonymously, did not prevent others from recording his munificence. Liberal offerings to Shearith Israel reached hundreds of dollars annually. In 1824 Hendricks accepted the honor of becoming the president of the congregation and held the post for three years, thus becoming the second Hendricks to hold a prominent office in Shearith Israel. Before his retirement from that office the congregation experienced a religio-ethnic problem for the first time in its history. The number of Jews accustomed to the German rite had increased and an influx of immigrants since the War of 1812 had doubled the Jewish population. It was clear that Shearith Israel could no longer meet the requirements of a heterogeneous population. As a result, after a series of conflicts and differences a new congregation was formed in the fall of 1825. Despite the fact that the members of the new congregation questioned the authority of the parent congregation by requesting that separate services be held according to their ritual, it was Hendricks, then president of Shearith Israel, who gave them the necessary support when they purchased property for a synagogue. The cost of the land and building was estimated at $8,300, and Hendricks advanced $5,000 toward this sum, which enabled the new congregation B'nai Jeshurun, to acquire its handsome Elm Street edifice. The five-year mortgage was offered at the remarkably low rate of interest of 1 percent per annum at a time when 7 percent was considered a reasonable rate, and 12 percent the usual rate. Two years later Hendricks cancelled the interest and deducted $150 which he had pledged as a gift to the congregation.

The rigid authority vested in the office of president of Shearith Israel was transcended by Hendricks in thus aiding a sister congregation, and this action revealed another phase of his character. He disdained personal reward, but was delighted when his son Henry was offered the honor of delivering the oration at the dedication ceremonies in the presence of the boards and members of both congregations. Not only were Shearith Israel and B'nai Jeshurun the recipients of Hendricks' benefactions, but within a few years the new congregation at New Orleans and B'nai Israel in Cincinnati were also to receive gifts from him.

His support of new literary endeavors was no less important. Hendricks encouraged Solomon Henry Jackson, the first publisher of Jewish books in the United States. As an author, Jackson gained his reputation by editing the publication of an antimissionary journal. When he later announced his intention to publish an interlinear Hebrew-English Bible in 1826 Hendricks became his first patron and gave the prospectus an official endorsement from the congregation. The publication of the Bible is not known to have occurred, but the encouragement given by Hendricks and the Tobiases did make possible the publication of Jackson's edition of the first Hebrew and English prayer book in the United States. During the 1830's, when Isaac Leeser of Philadelphia initiated a major book publishing program, Hendricks was prominent among the first subscribers. Throughout his lifetime Harmon Hendricks' interest in the publication of Jewish books never waned.

Personal taste brought Hendricks close to the fine arts. His younger daughters were encouraged to study art with Charles Canda, a New York drawing master who is not otherwise identified. Paintings and engravings graced the parlor walls of the Greenwich Street house, and alabaster statuary rested on the marble mantelpieces. As soon as the plan was announced he subscribed to John Vanderlyn's *Rotunda*, a panorama of city views exhibited in a specially constructed building that included Vanderlyn's own paintings. In his sponsorship of the new art center Hendricks was brought into association with such notables as Clement C. Moore, Cadwallader D. Colden, Mordecai M. Noah, David Hosack, Brockholst Livingston, and Valentine Mott. Hendricks also commissioned Samuel Waldo and William Jewett to paint a portrait of himself, which revealed the copper merchant as a pleasant but determined man.

Harmon Hendricks' interest in cultural and communal affairs left a permanent and positive influence on his children. Gradually, as his sons came of age, they were imbued with the same sense of philanthropic obligation, interest in Shearith Israel, and identification with the cultural and civic life of the city.

Much of this activity had brought them into contact with the Liverpool Tobiases, whose children were later to marry four Hendrickses. The Tobiases began to look toward America when peace was made between England and the United States. Writing to his brother Samuel, who arrived in New York from Halifax in 1815, Tobias I. Tobias informed him of Napoleon's surrender to Captain Maitland, of the scandal that busied the gossips of Liverpool when it was rumored that Dr. Samuel Solomon intended to marry a Christian lady, and of the state of the watch and chronometer business, some of which was not worth "egg broth." Colorful, warm, and as picturesque in speech as he was in his detailed letters, Tobias assured everyone, upon his dignity as an Englishman, that he would not settle in America. His curiosity, however, was aroused by the accounts of New York prosperity, and he finally succumbed to the desire to sail to the United States. He arrived in New York just at the time of the economic decline of 1816. Dry goods and patent lever watches were his stock in trade, but he could not compete with the flooded auction rooms that were undermining the economy or with the English hawkers who peddled a cheaper brand of watch on the city streets. Despite the difficult times three years of hard work were sufficiently rewarding to convince him that America was the land of the future, and in 1819 he returned to Liverpool to settle his affairs and convince his wife of the favorable business prospects offered in New York City.

By the time the Tobiases returned to New York Uriah Hendricks, whom Tobias envisioned as a prospective son-in-law, had left Columbia College and had gone to work at Soho. Under the tutelage of Sol Isaacs, who was popularly referred to in Jewish circles as "Steamboat Isaacs," Uriah became the third generation of the family to enter the copper business. Here he came in closer contact with his two cousins, Mathias Gomez, who began work at the same time as Uriah, and Aaron L. Gomez, who went to work at Belleville in 1820. Aaron L. Gomez had been a successful shopkeeper during the depressed years that followed the War of 1812, and because of his ability to weather this storm he was invited to enter the copper trade. Hendricks had come to value the benefits of attracting young men to the business, and it was not unusual for him to choose a bright and enterprising relative. Within the year the stylishly formal Gomez asked his employer for the hand of Harriet Hendricks. Consent was granted and in May, 1821, Gomez, "tired of a Bachelor's life," married Harriet, settled on the Soho estate, and succeeded Isaacs as overseer of the mill.

Immediately after Harriet's marriage Henry, having also left Columbia College, was introduced to the trade, joining his brother, brother-in-law, and cousin in learning the art of copper manufacture. At first Uriah and Henry concerned themselves with the mill's inner operation, the work of the mill hands, and the mechanical aspects of copper production. Mathias Gomez acted as a local sales and purchasing agent for copper and brass, and Aaron L. Gomez was responsible to Solomon Isaacs for the management of the mill. Uriah went to live on the Belleville estate in 1826, the year that he married mild-mannered Fanny Tobias. As copper production took an upward trend two younger brothers, Montague and Washington, were also admitted to the business. With these additions the Soho Copper Works was ready for reorganization.

V

In the spring of 1827 the term of partnership between Harmon Hendricks and Solomon Isaacs ended. A final division was made of the $171,000 in profits earned by Soho; the business was dissolved and immediately reorganized. Uriah and Henry Hendricks were admitted as partners and the younger brothers were designated as employees. Isaacs' name was eliminated from the firm name and the shorter title of Soho Copper Company was adopted. Profits were now to be shared differently: Solomon Isaacs was to receive one-third, and two-thirds were to be shared among Hendricks and his two older sons. The entire copper and metal inventory was valued at $319,955, and the new partnership was based on a three-year period.

The formal entrance of the Hendricks brothers into the copper business coincided with the most fascinating period of change in the early history of American copper. Many new uses for copper had been developed, and in three years more than 2,000,000 pounds of raw copper were manufactured at Soho. This broke all known previous production records. The introduction of illuminating gas in New York City created new uses for copper and brass, and chemical plants also began to employ copper utensils more widely. Pure copper, or caulk, found new markets. Brass, an alloy of copper and zinc, spelter, tin, lead, and antimony were in greater demand because of the rapid growth of the New England industries. Steam and ferryboat companies along the North River thrived as never before, and their consumption of copper exceeded previous records. A constant improvement of mechanical equipment by the engineering firms necessitated copper castings and special parts to carry out the designs of their ingenious mechanics. Robert L. Stevens took a leading part in developing the riverboat engine, James P. Allaire was pioneering in building compound engines, and the West Point Foundry Association was active in a variety of similar undertakings. All of these engineering companies depended upon Hendricks' ability to supply their special needs.

Throughout the twenties the effectiveness of copper sheathing was so well established that shipyards from New England to the South used it as a necessity. Shipowners considered it a good investment, and no sea captain worth his salt would sail in an unsheathed ship. The shipyards of New York were recoppering ships for the second and third time, and were never without a vessel on their stocks; they coppered at least one ship a week for Hendricks alone. Hendricks' contracts with the shipowners for recoppering were assigned to the shipyards. At the end of three, four, or five years, vessels returned to have their bottoms stripped and freshly sheathed. Some of the shipmasters bought copper from Isaac McKim if they were coming up from Baltimore, others bought from the Taunton Copper Works in Massachusetts, and a large number obtained English copper from New York metal importers. Then it was sent to Belleville for refining or manufacturing into sheets of various sizes and thicknesses. Five years before, this cooperation between the copper men would have been viewed as impossible. Now it was not unusual to have the starboard side of a ship sheathed with Soho copper and its port side with copper of another make. Hendricks, in his zeal to keep complete records, carefully noted these differences. Trade in copper had become so profitable in the favorable business climate that prevailed in the mid-1820's that competition was not a matter for concern. Manufacturing flourished and shipbuilding was encouraged by the increased trade. Concern was for quality and good workmanship, and Soho was happy to keep its mills rolling for the ship industry and others who preferred its work.

New Bedford and Nantucket yards wrote for copper to be sent up to them, or directed their ships to New York, where Hendricks personally appraised their coppering requirements.

Newport and New Bedford sailing masters and whalers bound for the South Seas, or engaged in the China trade, stopped at New York to have their ships boot coppered—nailing sheathing upon sheathing—to assure additional safety for their long and hazardous voyages. Trading vessels from the Argentine, packets from Havre, and passenger boats from Hamburg took advantage of their stay in New York and coppered their bottoms with Soho copper before returning to their home ports. New York's shipping merchants, a number that increased yearly, returned to Soho as regularly as the seasons, and the reappearance of such old accounts as the Quaker Hicks brothers, Samuel Thompson, Peter Harmony, and Lewis Wilcox occasioned no surprise. Young newcomers appeared on the scene to have their small craft coppered for the first time; for his new schooner, Seixas Nathan applied for both copper and credit. The aggressive and resourceful ferryboat captain, Cornelius Vanderbilt, who challenged the steamboat monopoly on the rivers of New York and New Jersey, returned to Hendricks regularly for copper and credit for his ship *Citizen*. Meanwhile a new breed of shipbuilders spread their yards along Corlears Hook, from where they could see the schooner *Two Fannys* glide by on its way to Belleville with a haul of copper. New York City now dominated the market in metals and determined their prices.

The purchasing agents, the commission merchants, and the metal scouts were busier than ever before. After thirty years of trading with Hendricks, Thomas K. Jones of Boston reached his highest peak in 1827, when the two men did a volume of business amounting to $46,000 annually. James Ward, who had reorganized his business under the firm name of Ward, Bartholomew and Brainard, had also increased his business with Hendricks. The Philadelphia agents were as alert to India or Banka tin from Malaysia as they were to the unconsigned shipments of Peruvian ore. The Baltimoreans Harrison and Sterret supplied every detail of information about the Maryland market that would be useful to their New York friend. Many new merchants were attracted to the expanding copper trade. Refineries and rolling mills were still few in number, but the center of copper manufacture for the middle Atlantic states remained in Belleville, New Jersey.

The sons of Harmon Hendricks were intrigued by the strides made by the copper trade, and its ramifications. They learned that much of the progress made by the shipyard owners had been due to the support and encouragement given by their father, who had helped Henry Eckford, the Wrights, the Browns, the Webbs, and Foreman Cheeseman to expand their businesses to keep pace with the rapid development of shipbuilding. The Wright Brothers had earned for themselves a fine reputation in general, and the respect of Hendricks in particular, because of their integrity and skill as mechanics. When they leased from Hendricks the half-acre lot on Third Avenue at Kips Bay on a yearly basis, no legal papers passed between the lessor and lessee; the agreement was oral. Foreman Cheeseman, with whom Hendricks virtually grew up in the copper trade, depended on his associate for a number of loans between 1821 and 1827 at rates of interest below the New York standard. Noah Brown's shipyard at Corlears Hook was mortgaged to Hendricks, and Brown was indebted to him for personal financial aid as well.

In the first generation of American shipbuilders the most flamboyant was Henry Eckford. He began his business shortly after Hendricks, and it was not long before he became a major innovator in ship architecture. When he returned to New York City after the war he sought Hendricks' support. Immediately after Eckford completed his work on the steamer *Robert Fulton* he was appointed contractor at the Brooklyn Navy Yard, and the copper that Eckford used for ships of the line came from Soho. Later Eckford grew dissatisfied with the manner in which the naval commissioners were supervising the shipbuilding program; he resigned from his position with the United States Navy to help develop the navies of Europe and South America. In order to carry out his plans he turned to Hendricks for funds and mortgaged his New York properties. Before the terms of the mortgage could be fulfilled Eckford died in Constantinople, where he had gone to organize a navy yard for the Turkish government.

With two of his sons as partners and two others learning the trade, Hendricks was able to devote more of his time to his extensive real estate holdings, to the discounting of notes, and to the banking and brokerage business. He hoped that the activity connected with these many occupations would not aggravate his rheumatism, which was becoming increasingly painful. His interest in the Hartford Bank, of which he was a director and a major stockholder, carried with it the responsibility of being its official New York representative. Bank notes were regularly supplied for exchange in New York, some carrying Hendricks' imprint; and private messengers traveled from Hartford as often as three times a week to deliver important banking instructions. The Hartford cashier who sent them was careful not to violate Hendricks' Sabbath observance, the more so when Hendricks became the representative for groups and individuals of diverse backgrounds. Connecticut merchants transacting business in New York did so from Hendricks' copper office, and in one case funds for a Montreal convent were dispensed through him. In advising Hendricks about the method of paying the convent's representative, James Ward did not refrain from writing of such "A novel transaction, a Jew banker for a Nunnery."

From Montreal inquiries of still another nature came from the Jewish Harts, an old Canadian family, who requested Hendricks in his capacity as a director of the Manhattan Fire Insurance Company to place the Hart fire insurance with his company. Hendricks was also involved in two other fire insurance companies, the Phoenix of New York in which he was a shareholder and the Hartford Fire Insurance Company, to whose financial aid he came when the Company was at a crucial juncture in its history.

Trade in stocks and bonds of every description had held Hendricks' attention since 1799, and eventually came to occupy a place of major importance in his holdings. Steamboat companies, bridges, ferryboats, government stocks, canal, and turnpike loans involved a large volume of his funds. Real estate investment, which had begun when the family first fled the yellow fever in 1798, exceeded $100,000 by 1830. Issuing mortgages was a lesser side activity. The church historian of old New York, Gabriel Disosway, tells of Hendricks that "he used to boast that in all of his immense money operations no one could accuse him of taking more than legal interest, and that in this respect he strictly kept the Law of Moses." Perhaps it was on this account that so many of the shipyard owners turned to Hendricks for financial aid since, as Disosway emphasizes, the practice of charging high interest rates was common among the businessmen of Wall and William Streets. Not only did the copper men turn to Hendricks for funds, but also a Livingston, and members of the well-to-do Moses family, and a host of small tradesmen as well. This was more than enough to keep Hendricks busy, while he directed Soho from the comfortable distance of his New York counting desk. By 1830 he had sharply reduced the correspondence with the English merchants that had once occupied so much of his time. He still maintained a correspondence with the well-known English banking firm of Baring Brothers for the negotiation of English and American stocks, and occasionally with Nathan M. Rothschild. The latter transactions were more for the transmission of cash for the impoverished Jews of Jerusalem than for matters of personal gain.

Another turning point in the history of the Soho Copper Company came less than a year after Henry Hendricks married Harriet Tobias. In April, 1830, the three-year partnership between Hendricks and his sons, and Isaacs was terminated. Meanwhile, Isaacs, seemingly a confirmed bachelor, had fallen in love with Elkalah Kursheedt and, at the age of forty-four, married the daughter of Isaac Baer Kursheedt, reputed to be a man of great Jewish learning. Subsequently Isaacs returned to live in New York, giving up his residence at Belleville.

When the firm was reorganized Isaacs and Hendricks, for reasons that are not made clear, concluded an association of three decades. Isaacs set up an independent business as a copper broker, was elected president of Shearith Israel, and entered upon an active community career. At the early age of sixty Harmon Hendricks retired, voluntarily relinquishing his position after having served the copper trade longer than anyone else in the United States. His sons, trading as Hendricks & Brothers, were launched as copper merchants of the third generation, prepared to serve a new generation of American industry.

5

COPPER IN AN AGE
OF TRANSITION

I

The early 1830's were as remarkable for the progress of the railroads as the twenties were for canals, turnpikes, and steamboats. Steam was applied to drive mills, to power the New England factories, and to turn the wheels of America's first locomotives. Greater speed and improved transportation were demanded by a mobile population. At the age of sixty Hendricks could recall with a smile the slower days when his goods were hauled by ox cart. Steamboats sailed majestically over the inland waters, packet boats were swifter, and the locomotive promised a new future for speed. While theorists were debating the superiority of canals over railroads, sparks from the first locomotives frightened the farmer through whose land the rails cut; and city folk, though eager to be among the first to board the trains, still gazed in amazement at the chugging vehicles.

Matthias W. Baldwin, the first major builder of locomotives in the United States was attracted to the magic of the steam engine soon after the English inventor George Stephenson had shown its successful operation. In New York the West Point Foundry Association attempted to duplicate Stephenson's English locomotive, but failed in its first undertaking. Other attempts followed and were more successful. Baldwin's now famous *Ironsides*, built in 1832, was one of these, and is remembered as the first of a series of locomotives that he manufactured during the

1830's. Copper was used for locomotive boilers, flues, and boxes, and was destined to make heavy demands on the rolling mills. The beginnings of the locomotive industry in the United States were as closely identified with Hendricks & Brothers as the steamboat and shipyards were with the work of Harmon Hendricks and the Soho Copper Company.

Uriah Hendricks could count eleven years of experience in copper when he assumed authority over the new company. Although his brothers shared the same financial interest, worked in harmony, and carefully divided their responsibilities, Uriah as the eldest son became the acknowledged head of the firm. In the first years of partnership major changes were avoided. A few minor innovations were introduced, such as the acceptance of printed business forms and a set of formal ledgers. Coal from the Pennsylvania mines and from the pits of Nova Scotia was tested for its quality; however, the largest quantity of coal for the Belleville furnaces was still shipped from Liverpool. The roads over which the coal was hauled on its way to the mill were newly graded. Improvements were made on the recently built

Uriah Hendricks, 1802–1869, son of Harmon Hendricks.

Morris Canal. Minor repairs were made on the mill houses and the adjacent living quarters were redecorated, but no structural changes were considered. The quarry was still worked with an eye for metal deposits and the surrounding lands were probed for copper ore.

Immediately after the reorganization was effected Uriah, his wife, and three little daughters returned to New York City. The New York offices, from which all of the business was conducted, also required the presence of Henry, who was now twenty-seven years old, and Washington, who had just turned twenty-four. Judging by the degree of his activity in the Jewish societies, Henry did not devote much time to the business, suggesting a reluctant interest in copper production. Washington was also attracted to the work of the Jewish charitable organizations and, oddly enough, he became a foreman of the local Fire Engine Company No. 20, despite his delicate health. When Uriah left for New York twenty-year-old Montague moved to the Belleville estate and joined his brother-in-law Aaron L. Gomez as an on-the-spot overseer.

Hendricks Copper Mill about 1860.

Uriah was responsible for the business records and attended to most of the correspondence, which was always in the name of Hendricks & Brothers. He informed old accounts of his father's retirement and assured them that "first quality American sheathing copper, a superior article of assorted sizes," would be supplied as theretofore. As new accounts were solicited there was a tone in the correspondence reminiscent of the letters that Harmon Hendricks had written when he assumed responsibility for his father's business thirty-three years before.

As American manufacture was developing, the demands for metal greatly increased. Iron production rose considerably after that industry was reorganized. Native ores enabled the forges and furnaces to work at full capacity, but the lack of sufficient American-mined copper kept its manufacture far behind that of iron. The two industries could hardly be compared as long as the bulk of copper used in the United States had to be imported.

To assure the firm a constant supply of copper many metal agents and commission merchants were alerted to the interests of Hendricks & Brothers. Thirty years had passed since Harmon Hendricks had employed John McCauley as his first Philadelphia agent. Solomon Moses, who had since become the official Philadelphia agent, continued in his former capacity, with copper receiving the greatest emphasis in his commission trade. Joseph Lyons Moss, the other Philadelphia agent, began to supply copper to locomotive builders and the engineering industries that had not previously been aware of its many uses. Moss's shrewd appraisal of the city's metal market proved to be of great value in negotiating important sales with the locomotive builders. Furthermore the valuable shipping and mercantile connections of the Moss family frequently enabled him to receive advance information about the arrival of metals at the port of Philadelphia.

Other Philadelphia commission merchants who dealt in metals —B. & I. Phillips, George Harley, or the Leaming Brothers— were as eager to sell their importations of Banka tin from Malaysia, spelter, or old brass to the New Jersey mill as they were anxious to purchase manufactured copper that could not be supplied locally by Nathan Trotter, the city's outstanding metal merchant. Francis M. Drexel, the miniaturist and portrait painter had already freed one hand from his brushes to negotiate the resale of imported copper to Hendricks & Brothers. Although Philadelphia declined in importance as a financial hub it remained an undisputed industrial center.

Other cities along the Atlantic coast were alive with the same activity. In Baltimore a new generation of copper men interested in the smelting and refining of copper entered into business; in New Jersey the cities that had lagged behind in manufacture saw an increase in the number of foundries; and in New England a large number of steam-operated factories were making extensive use of the malleable metal. New York State continued to be an important consumer, and the use of copper inland continued unabated. In New York City, close to the Beaver Street warehouse, uncle Sol Isaacs opened his own copper shop and resumed business with his nephews in the sale of brazier copper. From their own vast stock Hendricks & Brothers were able to offer 40,000 pounds of assorted braziery, raised and flat bottoms of standard sizes, and a large quantity of bolt copper. More than 20,000 pounds of Banka tin was constantly on hand, and composition nails were a standard article that could be purchased from them at all times.

The naval trade of the twenties had decreased, although the government still called for sundry copper supplies. Other dependable customers were the owners of merchant ships that navigated the waters of the New York area; farmers who depended on the use of copper stills for their grain; dye and chemical works that required vats; and the various waterworks—including the Manhattan Corporation. When the Hendricks brothers assumed responsibility for the firm they experienced the new demands upon the metal trade made by the revolution in industry. A gradual transformation in the uses of copper became evident: the call for raised and flat bottoms declined, and the manufacture of sheathing took an upward trend. The tin market, which had been dominated by Harmon Hendricks during the first two decades of the nineteenth century, underwent similar changes in procurement and distribution. Tin-plate competed with copper sheathing as a roofing material, and its other competitive uses attracted an enterprising group of metal merchants to the trade.

One of the new firms that influenced the tin market, and then in turn the copper market, and subsequently American copper mining, was Phelps, Dodge & Company. About 1812 the former saddle maker Anson Greene Phelps of Hartford, Connecticut, entered the New York trade, building up a business by selling tin-plate and brass utensils. With his first partner, Elisha Peck, Phelps was one of Hendricks' tin customers. When Harmon Hendricks shifted from the sale of tin-plate to concentrate on ship sheathing during the 1820's the tin market that he had dominated was slowly taken over by the energetic Anson Phelps. In 1834, when the new firm was organized under the name of Phelps, Dodge & Company, the name of Phelps was already synonymous with tin-plate. Hendricks & Brothers, who were still outstanding importers of bulk tin, had slipped into second place as retail distributors.

Although the company's emphasis on tin declined it never disappeared from the sales charts of Hendricks & Brothers. Other items in the nonferrous metal trade, spelter, lead, and antimony, received greater emphasis, but the most important field to which the Hendricks firm turned its attention was the production of boiler plate and copper strips for locomotives. This was a natural result of the revolution in transportation. Great strides were being made in the states of New York, Pennsylvania, and Maryland, where some of the first railroad companies were organized. Thirteen roads were chartered by the New York legislature between 1826 and 1831. The Mohawk and Hudson began to lay its tracks between Albany and Schenectady shortly after it was chartered in 1826. In 1828 the Pennsylvania legislature approved the building of the Main Line of Public Works, a system of interconnecting canals and railways that would link Philadelphia with the Ohio River. And in Maryland the Baltimore & Ohio Railroad was formally opened in the spring of 1830.

For their stationary engines and locomotives many companies obtained brass, composition, and boiler copper from Hendricks & Brothers. Among these were the New Jersey Transportation

Company, the Western Railroad Corporation of Massachusetts, the first companies of Michigan, and the many companies long since absorbed by major railroad lines.

Steam was no longer a theory when the stock of the Mohawk and Hudson Railroad, which at first could hardly be sold, became a desirable investment. It was possible for a New Yorker to board a steamboat for Albany, at which point connections could be made with the Mohawk and Hudson to travel the fifteen miles to Schenectady by rail. This seemed very practical and attractive to Harmon Hendricks, who bought more than 3,000 shares of the company's stock in 1833. New companies followed the same pattern of issuing stocks or bonds. As soon as the Utica and Schenectady, the Boston and Providence, and the New Jersey transportation companies opened their books for the sale of stocks Harmon Hendricks bought hundreds of shares in each, and encouraged his sons to do the same. Although he had retired from the copper trade he continued the investment practices that related to the trade in which his sons were now participants. His investments in the companies with whom Hendricks & Brothers were doing business in the 1830's were strikingly reminiscent of the investment practices of Harmon Hendricks with the steamboat and canal companies during the twenties.

At one time the elder Hendricks had also offered to underwrite most of the canal loan for the New York state commissioners, but this proposal was not accepted. In 1834, when a new loan of canal stock was floated, it was the banking firm of Joseph Brothers who arranged for a loan of $150,000 with the distinguished banker Nathan Mayer Rothschild. Wall Street gossipers whispered about a certain "Jew operator" who remained unnamed. The repercussions of this transaction were not felt until two years later.

While their father studied the transportation market Hendricks & Brothers added hundreds of new accounts, many of whose annual copper and brass purchases averaged approximately $1,000 each. Complete records for these years are not available, but some indication of the extent of their business can be had from the ledgers covering the end of 1833 to the early months of 1836, a peak year in copper sales. Baldwin purchased slightly more than $12,000; the United States Navy $6,358; the West Point Foundry Association bought copper castings, railroad plates, and boiler copper amounting to $6,000; and five local ferry companies obtained supplies that ranged from token orders to $2,400. In one six-month period, twenty-eight ships were coppered by Hendricks & Brothers and more than a dozen others were supplied with sundry copper items that amounted to thousands of dollars.

Henry Hendricks showed the least interest in the work of the New Jersey mill, and he became the firm's traveling agent to the South, acting as a bill collector, copper purchaser, and general representative. In Philadelphia, where he went to determine why Baldwin was in arrears, he was amazed to find the "establishment immense, covering I should think an *acre* of ground." Baltimore was next, and here Henry was successful in collecting $15,000 from Harrison and Sterret, who were also in arrears. He gave Baltimore considerable attention, because its port rivaled Philadelphia's and much of the South American copper was unloaded there for smelting before it reached New York for resale. This experience gave Henry an excellent opportunity to study the other markets, to absorb the details of the southern market and transmit them to the New York office. From Baltimore he traveled farther south, making contacts for the firm, renewing the De Leon family relationship, and meeting the Charleston Tobiases for the first time.

Reserved for Uriah Hendricks was the New England factory trade scattered along the Naugatuck and Housatonic Rivers of Connecticut: the Scovills of Waterbury, the Brainards of Hartford, the Crockers of Taunton, and the shipyards that spread along the coast from Maine to Massachusetts. The manufacture of copper and brass for the factories of the Connecticut Valley required techniques other than those used for the shipyards. Some of the metal went into the buttons of the nation's clothing, some was converted into molds for hatter's kettles, and a considerable amount was spun into the rollers that printed patterns onto calico cloth. The production of copper for these specialties was entirely different from the rolling of heavy sheets for the coppersmith, the ship chandler, and the locomotive builder.

Before the prosperous year of 1836 came to an end the mill machinery was again improved at a cost of $17,000 in order to meet the precision needs of copper tubing and other engineering demands. Copper was refined for reshipment to South America, and manufactured copper was exported to France for the first time in the Hendricks' history. The Hendricks brothers proved their ability beyond question. They understood the need to expand, to reinvest in the mill, and to change with the changing times. In the first five years of their management the total volume of business went beyond $250,000.

II

The rapid growth of the copper industry in the first years of his retirement must have been a matter of gratification to Harmon Hendricks. There was still greater satisfaction in the knowledge that his sons turned to the copper business with the same interest that he had. When he reflected upon the divergent paths of his nephews he could make interesting comparisons.

Uriah Hendricks Levy, who had determined to become an apothecary in boyhood, now owned one of the finest shops in the city, at Broadway and Murray Streets. Mathias Gomez, the bookseller's son, gave up the copper trade after ten active years and went to New Orleans, where he was killed in a duel in 1833. Uriah Hendricks Judah, in an attempt to emulate his cousin Benjamin, the controversial novelist, struggled to become a poet and then switched to journalism, but he depended upon the shipping trade for a livelihood. The De Leons were absorbed into the culture of Charleston and gradually drifted away from their New York cousins.

These descendants of the colonial Jewish families of New York represented an elite who associated more with Dutch and English Jewry, differed in religious observance from the German-Jewish immigrants, and developed a social outlook of their own. A few were secure in wealth, some in their cultural and communal achievements, others in their patriotic record in the American Revolution, but all were proud of being "old" Americans. Traces of this social outlook are found in the late eighteenth century among the Gomez, Seixas, Lopez, and Hendricks families. A generation later, when Harmon Hendricks' children reached the age of marriage, the trend became a pattern that was reinforced by business associations, marriage, and membership in Shearith Israel. The congregation and some of the Jewish societies it sponsored helped unite a number of the families where business ties did not. The Tobiases, comparative latecomers, with rich, vibrant personalities, expressing a great interest in Jewish affairs, immediately attracted attention and won the respect of their new associates. It was a source of pleasure to Harmon Hendricks when two of his sons married two Tobias sisters, and Montague chose Rachel Seixas Nathan for his bride. Less than a month after Montague's marriage in May, 1836, Emily Grace Hendricks married her new brother-in-law Benjamin Seixas Nathan, a sensitive man, and a promising young stockbroker. The Seixas and Nathan families, in addition to the Gomezes, were now all represented in the Hendricks family.

Their participation in the Jewish charities, still few in number and still limited in scope, was also on a family basis. One of the officers of the Society for the Education of Poor Children and Relief of Indigent Persons of the Jewish Persuasion, founded in 1827, was Tobias I. Tobias. Tobias induced his son-in-law Henry Hendricks to join the group of which he was treasurer. Six years later Hendricks became one of the managers along with his brother-in-law Benjamin Nathan. They were joined by Nathan's cousins, the Seixases, who were also engaged in the work of the society. Hendricks' intimate association with Nathan in this society led to the participation of both in the organization of the Hebrew Assistance Society. Nathan was its first president, uncle Naphtali Judah and Henry Hendricks were managers, and Henry Tobias, who was then courting Rosalie Hendricks, also joined the group.

Washington Hendricks showed considerable interest in all of these activities, especially in the Female Hebrew Benevolent Society, an organization patterned after the one founded in Philadelphia by Rebecca Gratz. His poor health, however, permitted little time for other interests. Montague, most of whose time was

Benjamin Nathan, from *Harper's Weekly*, August 20, 1870.

spent in New Jersey, could contribute only a financial share to the Jewish societies. Uriah contributed faithfully but was an active participant only in Shearith Israel. Of the four brothers, Henry Hendricks gave most liberally of his time to the many Jewish institutions that arose in New York City prior to the Civil War.

The times demanded more than the Jewish and other charities of New York were able to offer, because business conditions had once again become erratic. Early in 1837 New York was the scene of rioting by the unemployed; flour stores were raided, and what the hungry could not seize they destroyed. Before long other strata of the population were affected, and the nation was swept by bankruptcies. The Panic of 1819, well remembered by the elder Hendricks, was recalled by Uriah. Comparisons were made between the paper money issue of 1819 and the "rag money" of 1837. Hard money was scarce and paper money was worthless. The value of choice securities tumbled overnight, and Harmon Hendricks experienced heavy losses for the first time. When the railroads and the locomotive builders were unable to pay their notes Hendricks & Brothers had their first taste of national crisis.

In New York City the second American depression was attributed to the "spirit of wild speculation which had existed for the last two years." Silas Wright, the New York senator, wrote spitefully to President Van Buren that it was a cause for rejoicing when the Joseph Brothers, "Jew brokers of New York," failed. By April 11, 1837, the number of failures had reached 128, with over $60 million in liabilities. There were few Jews among them. The Hendrickses were hurt in their transportation stock, the value of which fell considerably. As the distress became widespread the banks of New York added to the disaster by the suspension of specie payment.

Echoes of these conditions quickly reached England, where the export trade had taken a sharp decline. Conditions in the two countries were discussed in the Hendricks-Tobias correspondence, where it was reported that English "goods were selling for the price of duty." From Liverpool came accounts of the English spinners who reluctantly joined the ranks of the unemployed. Buyers of Tobias lever watches were greatly reduced in number. Despite a knowledge of conditions in America Michael Tobias determined to sail for New York in the hope of stimulating sales of the watches and chronometers that he manufactured in Liverpool. He was attracted by the opportunity to visit his brothers, Samuel and Tobias, to meet the Hendricks family, and review the American market. Hardly a month after his arrival in New York, in the spring of 1838, Michael Tobias suddenly died. Since the body could not be returned to Liverpool for burial, Uriah Hendricks was called upon to make arrangements for a New York interment.

Before this sad news could reach the Tobiases in England Harmon Hendricks was taken ill. He had appeared in good health at the time Michael Tobias arrived in New York and, with the exception of his rheumatic complaints, he had been attending to his affairs in a leisurely and comfortable way. During the last week of March he was confined to his bed, and he failed to respond to medication. The newest electrical apparatus in medicine was used in an attempt to stir him out of a coma. When this, too, failed, the older treatment of leeching was applied, but without success. In the late evening of Monday, April 2, 1838, Harmon Hendricks, at the age of sixty-seven, died in the presence of his numerous family.

In his lifetime Hendricks was thought of as the wealthiest Jew in New York, but his significant contribution to American trade and industry had not been fully recognized. From his boyhood days as an apprentice up to the time of the War of 1812 Hendricks had been involved in every aspect of the carrying trade; he had experienced the restrictions of commerce in the West Indies and the British Orders in Council; his seamen had been impressed by the British, and he had suffered the effects of the embargo, the American tariff, and the English autocracy of the seas. Importation of copper, the business begun by his father before the American Revolution, became the backbone of a trade that included all sorts of goods manufactured in England and on the continent of Europe. When the important maritime use of copper was recognized Hendricks was among the first to further its application.

He was able to enter the business of refining and rolling copper when the War of 1812 had shut off the supply to the United States because his technical knowledge matched Revere's, his engineering skill paralleled Schuyler's and Roosevelt's, and his financial resources equaled Livingston's. The steamboat age, which changed the whole mode of American inland transportation by water and was identified with the first age of American copper, brought him into association with Robert Fulton, some of the first engineering companies, and the pioneer foundries of the nineteenth century. The shipyards of New York were even more indebted to him for his patience and foresight than for the funds and credit that he made available to their owners. He was a major force in producing copper for the expansion of the United

States Navy, which grew from a line of vessels that was almost beaten by the British to a large, organized fleet. The inland navigation companies of New York and Connecticut were equally dependent upon him for their growth. He was his own metallurgist at a time when the secrets of the science of refining metals were jealously guarded by the English. Above all he was a copper pioneer and a symbol in the rise of American industry.

No detailed record such as that which portrays the rise of Harmon Hendricks as a merchant and industrialist can be found for his personal life. The character of this man of few words and considerable reserve emerges only occasionally in his stray and fleeting observations of the business world, which he entered as a young man of twenty-three in 1792. His insight into business enabled him to detect the chicanery of an overseas agent but it did not always protect him from the colorful and scheming will-o'-the-wisp operations of a suave personality. He was not easily put upon but he was a man who could easily be appealed to. Yet he could drive a hard bargain with the shrewdest merchants. The thousands of invoices covering the period between 1798 and 1825, with their minute calculations of advance profits, in no manner suggest that he was a hard or parsimonious man. He was ethically bound to those friends and relatives to whose aid he came, like Isaac Gomez Jr., or Mordecai Gomez Wagg, although he knew all the time that his efforts could never turn them into practical men.

Hendricks must have been as convincing a man to others as he himself was convinced about the growing importance of copper. Behind his conviction lay his ability to influence his sons to join him in the work of the rolling mills. Their decision to enter the trade was somewhat different from the circumstances of voluntary choice that had led him, years before, to continue the business begun by his father. That it may have been a long-term plan for his sons is suggested by Hendricks' assignment of Soho shares in transportation stock in their names, his readiness to establish an equal interest in the mill for all four sons, and his generally encouraging attitude to young relatives who could form a responsible corps of hard-working men. The practice of involving male relatives in one business was typical of the times, but the idea was so deeply instilled in the sons of Harmon Hendricks that it surpassed any plan he might have devised.

At no time did Hendricks give any indication that he yearned for social position, desired to seek political office, or to strive for high status—goals that had begun to influence the wealthy of New York. To the contrary, he was content to spend his leisure time with his family and keeping active the traditional association with Congregation Shearith Israel.

If the merchants of New York knew little else of Harmon Hendricks they had no doubt about his religious principles. It was a matter of firm religious conviction not to charge usurious rates of interest; no merchant could tempt Hendricks to do business on the Sabbath; and the mill at Belleville was closed to all on Sabbaths and Jewish holidays. When these practices are considered, taking into account the background of a financial world that discounted notes at unusually high rates of interest, and the fact that Sunday laws reduced Hendricks' work week to five days, it was all the more remarkable that Hendricks maintained a high level of industrial production without sacrificing any of his religious principles.

Isaac B. Seixas, the spiritual leader of Shearith Israel, who officiated at Hendricks' funeral, was aware of this praiseworthy conduct, but this was of small importance to the bereaved family. Seixas' carriage followed the two-horse hearse that led a procession of thirty-six other carriages in what was the largest Jewish funeral witnessed in New York City up to that time. After the procession dispersed at the cemetery of Shearith Israel only the grave-watchers remained behind, in accordance with Jewish custom, to guard the grave of America's first Jewish industrialist.

The death of Harmon Hendricks brought Uriah, the oldest son, into a new sphere of activity. Uriah was appointed the administrator of his father's estate, which exceeded $1 million. Millionaires were so uncommon at this time that Hendricks was brought into the orbit of wealth associated with the names of Goelet, Schermerhorn, Rhinelander, and other New Yorkers. Although there were vast personal differences among these men and the methods by which they accumulated their fortunes, the number of millionaires at that time could be counted with ease. The estate and its final disposition was to keep Uriah Hendricks a busy man. As head of the family he was also sought out for counsel by relatives and close friends. Uriah's dependability in time of stress induced the relatives of Michael Tobias to seek his advice in administering the American affairs of their extensive watch and chronometer business, which was known throughout the English-speaking world. A reputation for devotion to his family and friends made him the unquestioned choice as advisor to his wife's relatives. Uriah had acquired much of his father's reserve, although his black hair, his medium height, and his enjoyment of laughter strongly reminded one of his grandfather.

Uriah Hendricks' relationship to the Tobiases was strengthened further when Henry Tobias announced his intention of marrying Rosalie, Uriah's younger sister. Her frail health did not detract from her charm, and from relatives in faraway Liverpool came the advice that the "balmy air of Broadway in the afternoon" might affect the hoped-for improvement in her health. In the fall of 1839 Henry and Rosalie were married, and again the mail carrying the news between New York and Liverpool was as regular as the sailing of a New York packet.

Henry Tobias could report that the "valued favors per Serius,

View of Broadway from Anthony Street, looking north, 1856.

Great Western, Virginian, & North American" arrived safely. To the families who sailed back and forth, the speed of the pioneer ocean steamers, the *Great Western* and the *Serius,* was as essential as information about their arrival, their departure, and the conditions that prevailed aboard ship. For those who adhered to Jewish food laws, "Dry bread, and rice for more than a fortnight" was not a gourmet's diet, but one that had to be observed. Charles Tobias, congratulating Henry on his choice of a bride, reported that the "Captain of the Steamer (Liverpool) . . . when he heard that Two 'Tobiases' were going out and that we were Jews went to Gill the butcher, and ordered choser [*kasher*] meat, now if this is true, it shows the man and regret that I am not one of his passengers . . ."

With such reassuring information a Liverpool trip for the newlyweds was recommended. When Rosalie became ill in Liverpool the trip was cut short, and in the spring of 1840, shortly after their return to New York, twenty-year-old Rosalie died. The grief caused by her death was hardly compensated for by the birth of her brother Uriah's eighth child, for now the family was alarmed by the state of Washington's health, which had worsened. In March, 1841, while Washington was confined to bed, Uriah's eight-month-old baby died, and two weeks later Washington joined his sister Rosalie in the seventeenth-century burying ground of Shearith Israel.

Frances Hendricks, the widow of Harmon, greatly saddened by four deaths in three years, was now even more dependent upon Uriah. More legal responsibilities devolved upon her oldest son who, in addition to settling the affairs of Rosalie and Washington, was still occupied with his father's estate. The copper business, which continued to be dull, was fortunately in the capable hands of Montague, Henry, and their brother-in-law, Aaron L. Gomez. The economic crisis that gripped the nation still had not seriously affected the Hendricks brothers.

III

The aftermath of the Panic of 1837 was felt in the development of American transportation for several years. A remarkable determination to overcome all obstacles, both financial and technical, characterized the growth of the railroads as they stretched from state to state. However the engineers, the mechanics, and the builders of locomotives were not spared by the disrupted economy. Recuperation was gradual because money remained scarce. Failures continued, and the Hendricks brothers countered the hard times as best they could. Credit was still hazardous and the bankrupts of 1837 struggled to recover from the disaster. Many of the Hendricks' accounts had not paid their bills for two and three years, and when their notes fell due they were unable to meet them.

"If my friends don't give me the time I want, I must make an assignment," wrote Matthias Baldwin, the locomotive manufacturer, in 1839. Forced into this awkward position of announcing that he was on the verge of bankruptcy, the forthright Baldwin appealed to Hendricks & Brothers for a letter that would express their faith in his ability to make good his debts. Such a letter, Baldwin hoped, would set an example to his other creditors, whom he planned to confront publicly, explaining his predicament to them. He was right. Hendricks & Brothers graciously consented to the appeal that helped to rescue Baldwin from bankruptcy. His firm was reorganized, and in the decades that followed an extensive business was carried on between Baldwin and Hendricks & Brothers. Considerable time and patience were required to prove such faith in Baldwin, who depended on the railroads to pay their bills before he could pay his.

The inability of the Michigan Railroad to pay the sum of $45,000, due jointly to Baldwin and Hendricks & Brothers, resulted from the failure of the Bank of the United States to meet its commitments. The plight of the Detroit and Pontiac Railroad was worse. One of its first locomotives was built by Baldwin with materials manufactured at Belleville. They, too, were unable to pay the note assigned to Baldwin and held by the Hendricks brothers. When the railroaders of Detroit were urged to make payment the chagrined officials offered to return the locomotive to Hendricks.

These obstacles did not prevent the Hendrickses from investing in the construction of locomotives for which there was no immediate order. Advancing funds for the building of individual

locomotives seemed quite an acceptable step in the effort to help
revive a staggering economy. Funds for this purpose were made
available first to Baldwin and then to William Norris, another
of Philadelphia's locomotive builders. Norris had also experienced
financial difficulties, and he had resolved them by assigning the
notes due him by the Reading Railroad to the account of Hendricks & Brothers. This arrangement enabled him to restore his
weakened credit and to improve his business facilities. Norris
was engaged in a rivalry with Baldwin, and the same astute care
that Harmon Hendricks had shown in avoiding involvement in
the controversy between Governor Aaron Ogden and Robert Livingston in the steamboat monopoly was now shown by his sons
in remaining aloof from the competition between the two locomotive builders.

Foreign interest in American locomotives was encouraged in
the early forties by the financier August Belmont, who obtained
a contract for Baldwin with the Austrian government for one
locomotive. Knowledge of this prompted Norris to write to Uriah
of other contracts: "I am happy to add that I have just received
an order from Berlin in Prussia for 26 locomotives." He had just
completed work on a locomotive for Hendricks & Brothers which
they were to use as a model of engineering for the industry, competing with a similar one that Baldwin had built. Asserting the
fine quality of his workmanship, Norris wrote: "We will guarantee this Engine to do more work, at less Expence, than any
Engine of same class of any other make in the world." Norris
was eager to have Hendricks & Brothers make known the work
of his shop to the New York and Massachusetts railroads, but
he asked the New Yorkers not to offer his locomotive to the
Western Railroad of Massachusetts, where Baldwin had just delivered three locomotives.

With increasing regularity the foreign railroads turned to the
United States for locomotives, bypassing the English industrial
centers for the progressive industries of Philadelphia. Norris built
for Hendricks & Brothers a large class locomotive, with tender,
copper firebox, and a boiler that contained 108 copper tubes
mounted on 6 wheels and weighing 30,000 pounds. This became
the model for the French, Danish, and Austrian governments,

who ordered a total of 70 locomotives, according to the report of Joseph Lyons Moss, Hendricks & Brothers' Philadelphia agent. All of the copper supplies for the foreign work came from New Jersey, and most of the funds for this undertaking were advanced by the Rothschilds. For a time Norris had purchased his copper for cash, but now that his credit was reestablished, he privately boasted of his success to Moss. All this Moss faithfully reported to the Hendricks' New York copper office, and in one of his letters he added the postscript, "I saw Belmont's letter to them yest'y—It was friendly and quite flattering. He must make a good deal out of them—he is striving to secure a large order for them from Cuba."

Less important than either Baldwin or Norris in the manufacture of locomotives, but no less indebted to the Hendricks firm, was George Harley, another Philadelphian. Harley's losses involved his credit standing. His continued inability to make good his indebtedness prompted Moss to recommend legal action. This proposal was rejected by Hendricks & Brothers, who adopted the principle established by their father of waiting for better times. In all of these instances the Hendrickses lost very little and ultimately gained a great deal. It is unfortunate that the financial records of this period, which could tell the complete story, have not survived in their entirety.

Increased demands for nonferrous metals brought about a decisive change in procurement practices. A revival of interest in South American copper, a reconsideration of the English tin market, and the reappearance of Silesian spelter contributed to the resurgence of the metal import trade and set the stage for sharper competition. Bars, pigs, and cakes of unconsigned copper began to arrive by the thousands from Coquimbo, Valparaiso, and Lima. The port of Philadelphia received the largest quantities, with much of it appearing on the open market there. These conditions gave the Philadelphia agents the opportunity to show their acumen. Moss, sensitive to the choice copper and aware of each shipping transaction because of his excellent contacts, immediately sought advice from the Hendrickses. If Moss was absent from the city Solomon Moses was sure to be present to negotiate an important purchase.

The Taunton Copper Works belatedly learned the value of having a Philadelphia agent and finally sent a representative there. Acting on behalf of Baltimore merchants were the Ettings, an outstanding family of Philadelphia merchants who were tied to the Baltimore trade, where another branch of the family had interests in the Baltimore & Ohio Railroad. Their previous association with Harmon Hendricks was not continued by his sons. The close connection that developed between the Mosses and the Ettings, both socially and in the synagogue, did not lessen the competition for metals.

The Philadelphia trade had to be constantly on the alert to meet competition coming from Nathan Trotter, who by this time surpassed all others in the city in the importation of metals. The importance of the Philadelphia market could not be overestimated, yet though many tons of copper passed through its port and the city could boast the best industrial works and some of the finest factories in the country, its manufactured copper had to be obtained from either Baltimore or from the New Jersey mill. It was no secret, wrote Joseph Moss to Uriah Hendricks, that the copper men of Philadelphia "say your mill is always going & that you can furnish more expeditiously than any other in the Union." During the late 1840's even Trotter returned to Belleville for manufactured copper. The good reputation of Hendricks & Brothers was by no means exaggerated.

Nonetheless it must have come as a shock to Uriah Hendricks when his brother-in-law Charles Tobias advised him that Stark, Day, Stauffer and Company, an outstanding New Orleans firm, had not heard of Hendricks & Brothers! Instead they purchased English copper from Phelps, Dodge & Company of New York. Tobias described the New Orleans market as a major consumer of copper and urged a personal visit to the city. The New Orleans trade had an estimated 1,000 steamers engaged in a vast river traffic, feeding the rapidly growing Midwest on both sides of the Mississippi River. Vats were required by the sugar refineries of Louisiana, stills were in demand by the grain merchants of Kentucky, and sheets of copper were needed by the German smiths along the banks of the Ohio River. Almost a half-century before, Harmon Hendricks had supplied the New Orleans trade through

Jacob Hart, and now that trade had spread from the old South into the newly carved states and the sprawling territory of the Southwest. Texas had become a subject for conversation. Debates in Congress concerned themselves with the new republic. "Both Van [Buren] and Clay have come out against annexation," wrote Henry Hendricks to one of the Tobiases. Meanwhile the New Orleans business quarters used for the sale of Tobias lever watches also became an office for the copper merchants of the North.

IV

The mid-forties were more fabulous for the fundamental changes that influenced the copper trade than any previous period of the nineteenth century. Once a vast carrying trade had brought most of America's copper from England; this trade was disrupted by the War of 1812 and became dependent upon importations from South America. Then impetus was given to home manufacture, and a renewed interest in exploring beneath the surface of the earth for every variety of ore became typical of the period. A school of American geologists came into existence, and a number of states officially encouraged geological testing. Years of uninterrupted search for native ores finally proved successful. Reports appeared of Connecticut Yankees digging into the bowels of Hartford County for copper ore. The search for copper in New Jersey proved to be a disheartening record of failures: The Schuyler mine, worked at intervals, had already yielded its best ore, and optimism about nearby areas was never fully substantiated.

It was not until the late thirties and early forties that exploitation of the copper deposits of the Lake Superior region, known since the middle of the seventeenth century, was seriously undertaken. There is no fixed time for the beginnings of copper mining in this area, the Keweenaw Peninsula, but as soon as the Chippewa Indians were dispossessed of their title in 1843 mining permits were issued to the Pittsburgh and Boston Company and other investment companies. The long road forward, from the

Indian arrowhead, from chance individual discoveries by priests, trappers, and prospectors, had come to an end; the planned organization of mining companies had begun. Geologist, capitalist, and miner sought to exploit the reddish wealth. Hendricks & Brothers were aware of these copper lands and had obtained large tracts through purchase from the Public Lands Office prior to 1842. At first the scant output hardly sufficed to meet the nation's requirements, but as the yearly yield increased speculators and investors poured millions into the exploitation of the mines before even one cent of copper could be retrieved. Within a decade the Isle Royale copper lode near Houghton, Michigan, was found to be productive, and the Stock Exchange of New York and the brokerage firms of Boston, stimulated by the discovery of California gold, plunged deeply into mining securities of various sorts although, according to one historian, "copper not gold" became the leading Wall Street fancy. And so a new chapter opened in the development of the American copper industry.

Michigan copper mining eventually served to attract the heavy industries from New York and New England to Pittsburgh. The cost of transportation of ore to Boston was $15 per ton. By shipping it across Lake Erie to Cleveland and then to Pittsburgh, where the Hussey works for the smelting of copper were established, transportation costs were cut in half.

The Hendricks brothers continued to send finished copper from the New Jersey mill to Cincinnati by way of Pittsburgh by canal and riverboat, at the rate of 75¢ per 100 pounds. Unperturbed by the likelihood of losing some of their New Orleans customers to the Pittsburgh mill or by the manipulations of the copper speculators, they were more concerned with the effects of the disastrous fire that swept through the commercial district of New York in the summer of 1845. The Hanover Square area, where their grandfather had first entered business, was almost entirely ruined. Tobias I. Tobias' mercantile house was miraculously skirted by the flames, but the Hendricks' warehouse, though it narrowly escaped destruction, was damaged. The whole area had to be rebuilt at a time when the value of New York real estate had fallen and the carpenter, the stone mason, and the laboring man were seeking greater compensation for their work.

Despite the problems caused by the fire, the rebuilding plans, and the dull real estate market, the Hendricks brothers were as attentive to their families as if they had no problems at all. Joshua, Uriah's oldest son, was preparing for his bar mitzvah; Isaac, the son of Henry, had just celebrated his, and grandfather Tobias was as happy in the anticipation of Joshua's as he had been in drinking a toast to Isaac's. Mail between New York and Liverpool, conveying news of the celebration and good wishes to the youngsters, was more frequent than ever. The gift of a prayer shawl to Joshua by grandfather Tobias was accompanied by his blessing and the wish that Joshua, too, might "experience the same pleasure and gratification I do when you present a like symbol to your grandson."

Between family functions and religious festivities Uriah Hendricks finally determined to move ahead with a public sale of his father's real estate. Extensive holdings of valuable building lots and improved property in the first, fifth, fifteenth, and sixteenth wards were only a segment of the larger holdings that had been offered at private sale, but the poor real estate market attracted few buyers. Some real estate appraisers claimed that the property had soared to a value of $12 million. Auctioneers, vying for the opportunity to knock down the choice lots on Wall Street, Greenwich Street, Hudson Street, Trinity Place, or Exchange Place, were unable to convince Uriah Hendricks of their skill at the block. A number of them appealed personally to the warmhearted elder Tobias in the hope that he would intercede with his reserved son-in-law, Uriah. Amused by these proceedings, the Tobias women jotted them down for their Liverpool relatives to read. One can only speculate on the behind-the-scenes conversations before Anthony J. Bleecker was chosen to cry the sale.

The springlike winter of January, 1846, is better recalled for its hard times than for its pleasant weather. There was talk of war in the Southwest, and in the proclamations that argued for peace. Stocks rose one day and slumped the next; hard money was scarce, and immigrants from Europe, unaware of the difficulties that were to confront them, streamed into New York and crowded into the lower section of Manhattan. On January 29, when the cold pinch of winter was again felt, a curious crowd, including a number of the Hendricks family, attended the Merchant's Exchange to hear the auctioneer cry a sale that carried with it a variety of building restrictions. The sale met with a poor response. Benjamin Nathan made some purchases and a few parcels of the unsold lots were divided among interested members of the family. However the bulk of the real estate was not sold. If any of the properties had been eyed with envy there were no buyers to prove it, and some of Manhattan's choicest land waited another generation before it changed hands.

Later that year the offices of Hendricks & Brothers at 32 William Street were remodeled. The quarters at 77–79 Broad Street were leveled and replaced by a spacious four-story brick warehouse. Although the company was prepared to undertake the work of the United States Navy, which had meanwhile resumed its building program, every phase of the firm's activity was marked by growth and expansion in this period of the late forties. In addition to the Brooklyn Navy Yard's contracts for tin and brazier copper, the Navy had begun the construction of steamers. Much of the engine work for the Navy steamers was contracted for by the Philadelphia firm of Merrick and Towne, whose Southwark Foundry was distinguished for producing the fine heavy machinery required by propeller steamers. All of the extensive copper needs were subcontracted to Hendricks & Brothers, who guaranteed to meet government inspection requirements. The screw war-steamer *Mississippi*, later a Civil War monitor, and the *San Jacinto*, which was to be involved in the Trent affair, were both outfitted by Hendricks & Brothers in the new naval program. But not without difficulties. The hard-rolled copper for these vessels would not bear the flanching that was expected of it, and most of the copper had to be specially rolled in order to obtain the necessary strength without being too soft or too brittle. These technical problems were overcome at the Hendricks brothers' expense. At last the engineers were satisfied, the work was approved by the naval examiners, and a standard was set for other propeller frigates that were then under construction.

Between 1847 and the years immediately preceding the Civil War, a more serious gap in the records of the mill prevents even a speculative appraisal of its activity. Only a miscellaneous collection of mining, steamship, and railroad stock certificates which remains show the lively interest that Hendricks & Brothers maintained in the investment field. To this may be added their interest in the Belleville White Lead Company and in the manufacture of rosin oil by the New Jersey Oil Company. But the extent of their manufacturing, their interest in Michigan mining lands, and their methods of procurement and distribution of metals cannot be accurately reconstructed.

Shares in the New York and Havre Steam Navigation Company confirmed their continued interest in navigation, which now extended itself to overseas companies; stocks in the Kentucky and Mississippi railroads showed that their attention was not limited to the railroads of the East; and mining securities, which were vying with desirable railroad stocks, held the greatest interest for the copper firm. The promise of Tennessee copper induced Uriah Hendricks to purchase more than three hundred $100 shares, but nature was not to fulfill the dream of the Sewanee Mining Company of Tennessee or other southern gold and mining companies.

Lake Superior stocks competed sharply with those offered by the southern companies and, because the copper yield of the new mines was still only promising, it was difficult to determine the desirability of one over another. Issues of unauthorized stocks by the new mining companies and by the banks and railroads confused the unwary purchaser and the novice. As a result, investment became a highly involved, speculative, and unpredictable gamble. During 1855 and 1856 the stock market showed an upward trend, but the following year it fluctuated wildly and collapsed in one paroxysm.

A general panic swept Wall Street, credit disappeared, and the hard times of 1837 were again repeated. From their William Street office the Hendricks brothers could see excited men filling the streets on their way to Wall Street. Work stopped and the laboring man attempted to withdraw his savings from the New York banks, eighteen of which hurriedly suspended operations.

In October, 1857, during the height of the panic, optimistic mining speculators determined to organize a mining board, in offices opposite those of the Hendricks brothers. The mining board met the same fate as the stocks it hoped to control, and within a few months it came to an end.

Hendricks & Brothers suffered losses with their mining stock, but the extent cannot be evaluated. The will-o'-the-wisp copper shares in the various southern companies were filed away in hopes of a better future. The structural soundness of the firm saw them through another economic crisis. Although it would be difficult to determine what their business views were at this time, or the volume of their business, enough correspondence survives to show that the three brothers attended to their affairs with concern but without alarm.

V

In 1857 Uriah Hendricks was the father of sixteen children, all of whom were to survive him; Henry was the father of eight, and Montague the father of six, three of his children having died in infancy. Of this extensive group of thirty cousins, most of whom were girls, Uriah's four sons were the first to come of age. Joshua, the oldest, was eager to enter business but had not yet considered the copper trade. At twenty he was sent by his father on a tour of New England, during which he described "the follies of the Fashionable world," giving his father the opportunity to reply that, "like the tinsel of stage decoration it all looks like gold." Joshua then traveled south to visit the Charleston Tobiases, one of whom, Thomas Jefferson Tobias, had married his sister Adelaide. This Tobias, unrelated to the Liverpool family, was anxious to come north and join Joshua in the business of commission merchants. Tobias was also eager to learn more about the copper trade, but for the time being nothing came of this interest and the two young men pursued their own ways. Joshua became an agent for the Jewish firm of Pereyra Freres of Bordeaux, distributing their fine brandies in New York and New England.

Early in 1857 Joshua married Emma Brandon, thereby bringing together two families known to each other for four generations. The Brandons were related to the Jamaica Correas and
the Charleston De Leons. Emma's father, Joseph, had settled in
New York in the 1820's and had become a stockbroker, a specu-
lator in western lands and a close friend of Uriah Hendricks.
Few New Yorkers were as informed about railroad and mining
stock as Brandon, and fewer still had been fortunate enough to
survive two economic depressions in two decades. The knowledge of the stock market that Joshua gradually acquired un-
doubtedly came from his father-in-law.

While Joshua was courting Emma Brandon his brother Francis,
two years younger, was completing his studies at Columbia College. Francis, despite the two academic degrees he had earned
at Columbia, had no intention of becoming a professional, but
looked forward to the exciting world of business. Edmund was
determined to emulate his older brothers, and Harmon Washington, the youngest of the four, was still unconcerned about his
future.

Of Henry's children, Isaac was the oldest, and at the age of
twenty-six he was already an active importer. Copper attracted
him not at all. His only brother, Arthur, was still a boy, who
later chose to become a physician. Montague's two sons, Mortimer and Harmon, also planned to enter trade, and both of them
were to have the benefit of their parent's experience and connections.

The traditional closeness of this large family in so many aspects
of their activity is remarkable in American Jewish life. The
descendants of the first Uriah Hendricks, the Levys, the Judahs,
and the New York Gomezes, living in the same city, active in
the same Jewish congregation, and engaged in allied trades,
were drawn closer from decade to decade. Time gathered up
the generation contemporary with Harmon Hendricks. Naphtali
Judah, the last of the eighteenth-century group, died in 1855
at the age of eighty-one. His "Lucky Lottery Office" was still
recalled by many, long after lotteries had passed out of vogue,
and the differences he had with Harmon Hendricks were long
ago forgotten. But the family ties between the Judahs and Hen-

drickses remained. The younger generation, the fourth in the Hendricks line, were to extend this elite link into the twentieth century.

In the fourth and fifth decades of the nineteenth century the immigration from Germany, which had been a trickle in Harmon Hendricks' time, had turned into a steady stream. Downtown Manhattan was enriched by the colorful habits and the enterprising manners of the newcomers, who crowded into Chatham and Houston Streets. The Jewish population of the area increased tremendously. At one time counted in the thousands, New York's Jews could now be reckoned in the tens of thousands. Where there had been one synagogue in the early twenties, the area now had twenty; and where there had been two or three charitable societies, there were now almost thirty. It was a homogeneous group only in the eyes of the stranger, for the immigrants were separated by linguistic differences, divided geographically by neighborhoods, sustained by a small-town or rural outlook, and they practiced a peculiarly different mode of worship, all of which removed them socially from the older families.

One of the problems that concerned the Jews of New York was the lack of medical facilities under Jewish auspices. For a number of years this had been a subject for discussion, and influential members of the community had been urged to take action. Through the efforts of the Hebrew Benevolent Society, Henry Hendricks and Theodore J. Seixas issued a call for the founding of a hospital. Finally, in 1852, Sampson Simson and eight others stepped forward to organize the Jews' Hospital. Hendricks was joined in this major undertaking by his brother-in-law Benjamin Nathan, and they became two of the hospital's nine trustees. Hendricks was the first treasurer and Nathan the first secretary of the hospital. For the first few years following its organization the active involvement of the Hendricks family and their inspiring gifts contributed much to the hospital's ultimate success. The alliance of the Hendrickses with others, the uniting of foreign-born Jews with the older families in a common cause, was successful. Support for the movement to organize the hospital, and the leadership provided by the Spanish-Portuguese Congregation Shearith Israel came in large measures from the Hendricks family.

The same quality of devotion to Jewish communal work that

was typical of Henry Hendricks was characteristic of his son Isaac. Isaac assumed the responsibility for raising funds to build a synagogue in Geneva, Switzerland, at a time when the Swiss Treaty with the United States—restricting the travel of American Jews in Swiss cantons—was hotly debated. Other members of the family participated in various aspects of overseas aid. Through the offices of Sir Moses Montefiore the daughters of Harmon Hendricks, Hannah, Hermione, and Selina, transmitted their gifts for the aid of Jews in Palestine and the poor of Jerusalem. Occasionally, for the sake of convenience, they used the exchanges of the copper importers Phelps, Dodge & Company to expedite these funds. Selina, whose time was occupied in teaching the arts of ornamental needlework, singing, the pianoforte, and the guitar, also managed her own gift-giving. She contributed funds to the Jews of Lancaster, Pennsylvania, for the restoration of the colonial Jewish cemetery of that city, and in addition supported the efforts of the newcomers in building a synagogue in Lancaster. During the mid-fifties, when the widespread growth of relief agencies revealed the true state of economic conditions, when money was scarce and contributors few, gifts such as hers of $50 or $100 were indeed unusual.

Isaac Leeser, the editor of the *Occident*, the foremost Jewish journal of its time, was a severe critic of the older Jewish families of America for their aloofness, their Spanish hidalgo attitude and for their penny-pinching. He singled out the liberality of the Hendricks family as an exception. Privately Leeser took the liberty of recommending needy institutions or societies that he believed warranted special support to members of the Hendricks family. Modesty and reserve continued to govern the family attitude in matters of philanthropy, and the practice of keeping such activity from the public eye, begun by Harmon Hendricks, was maintained by his descendants. Again, only the records of the recipient reveal the conduct of the giver.

VI

In 1859 one of the symptoms of gradual recovery from the panic was the organization of a new mining board. Immediately after its organization it opened offices at 24 William Street, closer still to the Hendricks office. This time many of its members were drawn from the New York Exchange Company. However, the Hendrickses remained aloof, watching the speculators vie with the serious investors in riding the tide of popular investment. As the economic recovery continued the younger men of the family embarked on careers of their own. Joshua resolved to enter the copper business; his younger brother Francis accepted the offer of Thomas Jefferson Tobias that had previously been rejected by Joshua, and the two became commission merchants, forming the firm of Tobias, Hendricks and Company. Their cousin Isaac joined with a brother-in-law, Henry S. Henry, and Solomon De Cordova in organizing a mercantile house that had connections from Liverpool to the heart of Texas. Mortimer Hendricks, a son of Montague, and a trustee of Shearith Israel, entered into a similar type of business on a less ambitious scale. The new generation of Hendrickses took the first step away from the copper trade and established a business elite of their own, which included that branch of the Tobias family who were among the earliest Jewish settlers of Charleston and the Jewish De Cordovas, pioneers in Texas land development.

During 1859 the Hendricks brothers, strong advocates of steam,

Belleville Copper Rolling Mills, about 1870.

Edmund Hendricks, 1834–1909.

belatedly introduced it into their own mill. Before that innovation, waters coming from the Passaic River had powered the mill, but a reduction in the water supply and the increased production of copper made the use of steam a necessity. Two steam engines, one of 300 and one 90 horsepower, were added to the mill facilities. The cost of this and other mill improvements, in addition to modernizing the properties on the estate, approximated $50,000.

In New England the copper trade was able to report a growth and expansion. The Swifts of New Bedford, who reaped much benefit from New England shipping by repairing the whalers, clippers, and schooners, determined to build a copper works of their own. For over two generations they had bought copper from the Hendricks family, but this was no longer practical. The increased volume of shipping and the expansion of the whaling industry made it imperative to have a rolling mill of their own that would be within easy distance of the shipyards of the old whaling port of New Bedford. In 1860 they organized the New Bedford Copper Company, with a capital stock of $100,000. The Hendrickses held the Swifts in such esteem that they immediately subscribed to a number of shares in the undertaking of their former customers, now potential competitors. The shrewd Crockers of Taunton, who had recently enlarged their own works and had opened a refinery in Rhode Island, viewed the project of the enterprising Swifts with undisguised concern.

By the time that Abraham Lincoln was nominated for the Presidency copper in America had risen to a position of importance second only to the vast iron industry. American ores came into the market from the Lake Superior region in increasing quantity. Copper mined by the Norwich and the Isle Royale Mining Companies and others of the Michigan area reached the East by water transportation across the Great Lakes system, and both production and transportation gave stimulus to the industries of the East.

VII

In November of 1860 the state of South Carolina, provoked by decades of dissent and the election of Lincoln, determined to declare herself a free and independent commonwealth. Few were ready to believe that this secessionist stroke, which was a culmination of differences about states' rights, a conflict between southern slavery and northern profits, and of a long history involving the economic practices of the Americans, would erupt in a destructive war. That it would bring about a complete revolution in the metal trade could not have been envisioned by the most foresighted man in the industry.

The War between the States followed the South Carolina Ordinance of Secession, the disaster at Fort Sumter and the organization of the Confederate States of America. War production utilized all the northern industries, particularly the newly built and improved copper refineries and rolling mills. It was as if fate had destined their readiness and preparation.

Among the first casualties of the war were the northern businesses that had one foot in the South. Half of the interests of Tobias, Hendricks and Company were in Charleston, the seat of the secessionists, where Tobias' brother Joseph was responsible for the southern half of the firm. Before the echoes of the guns that subdued Fort Sumter faded over the Charleston harbor, and before the northern blockade of southern ports became effective, hasty preparations were made to transfer the Charleston assets of the firm to Cuba and thence to New York. Much was lost when the business was disrupted, and the stock coupons from the Florida Railroad, payable in the spring of 1861, soon became a valueless curiosity.

The firm was quickly reorganized to meet the crisis. Thomas Jefferson Tobias, torn between southern loyalties and a northern partnership with his relatives, became remarkably silent about the war. Secessionist sympathies also divided the partnership between Isaac Hendricks and the Texas De Cordovas, and only a northern commission business was continued for the duration of the war.

Unforeseen changes during the first year of the war also confronted the copper firm. In the spring of 1861, when the board of the Jews' Hospital was discussing the advisability of making available a ward for the use of the Union Army, Henry Hendricks resigned his office as treasurer because of poor health. Before an affirmative decision could be reached about the ward Henry died at the age of fifty-seven. The business now went to Uriah and Montague, with Joshua taking a more active part in all of its affairs. The only change made was that the firm adopted the new name of Hendricks Brothers.

Wartime consumption of copper by the Army and Navy exceeded all other previous demands. The production of ship and locomotive copper, ordnance, and copper ingots, reached its peak in the history of nonferrous metals. The New Bedford copper mill completed its building in time to be of advantage to the Union; the improved facilities and the introduction of steam at

Belleville placed Hendricks Brothers in a unique position for wartime production; the Reveres and other copper firms throughout the nation mobilized all of their resources for the vital war work, which had to be dispatched with unprecedented speed.

A shortage of skilled labor quickly became evident when men responded to the call to arms. Edmund and Harmon Washington, the youngest of Uriah's sons, were ready to enter the copper trade, but Edmund instead followed his cousins, Harmon and Frederick Nathan, and joined the Seventh Regiment New York National Guard. Edmund never reached the field of battle because a throat injury sustained toward the end of 1861 exempted him from active duty. Harmon Washington, only seventeen, divided his time between learning the elements of the copper trade and continuing his education.

At the same time their father was urged to become active in the Central Finance Commission of the New York Sanitary Commission to help raise funds for the relief of the military. It was one of the few civic posts that he accepted, and only his strong Republican views overcame the retiring attitude which normally

Dam at Hendricks Brothers Copper Mill, Belleville, New Jersey, about 1870.

BELLEVILLE COPPER ROLLING MILLS.
SOHO. NEW JERSEY.

Hendricks Brothers
49 Cliff Street,
New York

The mill after it was rebuilt in 1874.

compelled him to bypass other honors of public office. The women of the family organized themselves into the Ladies Army Relief group to sew clothing and roll bandages under the sponsorship of Congregation Shearith Israel.

After Edmund returned to civilian life Hendricks Brothers sent him on a tour of the Midwest to solicit work from smaller companies. All of the energy and zeal that might have gone into the field of combat he now turned to the copper business. Unlike his brother Joshua, Edmund did not have a good opportunity to learn the trade. The conditions of the war had prevented him from studying the mill operations or obtaining office experience. His knowledge of the trade pertained more to prices, the availability of metals, and wartime work. While he was on the road a constant correspondence with the New York office kept him informed of the scarcity of copper, the demand for lead, and the fluctuating price of brass. Prices rose with the news of each battle and metals became scarcer with each government contract. Spelter, especially the Silesian imports, was more widely used than before. Tin and tin-plate were once again costly and lead was more expensive than ever.

Edmund's resourceful knowledge of transportation facilities was helpful in expediting supplies from mine to mill and from mill to military depot. The ability of the railroads to bring copper to the East without diverting their equipment from troop movements was of primary importance. The strap-iron tracks that had been laid during the fifties, reaching sparsely settled areas, now carried raw materials to the industries as well as finished products for war to the field of battle.

Railroading and locomotive building underwent decisive changes during the first years of the war as transportation became an effective weapon in the Union strategy. To meet military requirements, the Baldwin Locomotive Works of New York gave Hendricks Brothers heavy orders. In Philadelphia John Edgar Thomson, the president of the Pennsylvania Railroad and a famous figure in early railroading, showed his friendship for the Hendrickses by introducing Edmund to the subsidiary branches as a member of "the great copper house" of New York. The Pennsylvania's lines connected with a flourishing system that led to the South and the West, and Thomson was eager to have the Hendrickses obtain as much of his company's work as possible. Other lines in which the Hendrickses were interested developed during the war years. In 1863, when the Atlantic and Great Western Railway Company opened its broad-gauge connection, Hendricks Brothers was invited to participate in the ceremony because of the company's involvement in the railroad industry. Over many of these new roads the copper and brass that was manufactured at the Belleville rolling mills went to the armories and arsenals of the North.

The war not only mobilized the railroads but also affected the entire copper import trade. Copper and tin from England and spelter from Silesia arrived in the port of New York on British bottoms. Much of American shipping had to be transferred to British merchantmen to avoid seizure by the Confederates, while Union vessels were mobilized for the blockade against the South.

As the war continued, the South, battered by land, retaliated on the waters. The Confederates had raised the abandoned frigate *Merrimac,* rebuilt her, and covered her sides with four inches of iron plate. Then she steamed out to meet the *Congress* and the *Cumberland* and, with her impenetrable ironclad sides, from which Union shot fell like pebbles, she sank the *Cumber-*

land and accepted the surrender of the *Congress*. Returning the following day to destroy other ships of the Union Navy the *Merrimac* encountered her match.

The Union Navy had been aware of the Confederate plans for rebuilding the *Merrimac* and had embarked on a program of its own by accepting the offer of John Ericson to build an ironclad. On March 9, 1862, the Brooklyn-built *Monitor* arrived in time to meet the *Merrimac* at Cape Henry. After hours of fighting, both ironclads withdrew, neither inflicting serious damage upon the other. This famous battle between the two ironclads marked the end of the era of wooden ships and the use of copper for sheathing wooden hulls. Even though it did not immediately affect shipbuilding it was to have a decisive influence on the American copper industry and would change the naval architecture of the world.

Yet for the time being ship coppering continued as before, and the shipyards of New York could not recall such an abundance of work in their history. Vessels were being altered to suit the Navy Department for the blockading fleet, steam transports were under constant construction, and the builders of wooden hulls so far had ignored the impact of the ironclads. The yards of New York were so crowded that Joshua Hendricks was compelled to turn to Neafie, Levy & Company, the Philadelphia shipbuilders, to learn whether they could build two lightships in which the Hendrickses were interested.

Mining, banking, and brokerage continued to occupy much of Hendricks Brothers' time. The promotion of gold and copper mining stocks was in no way diminished during the war years. In fact they competed strongly with the stocks that were issued for the support of the various New York railroads, in all of which the Hendrickses had an interest—either as stockholders or as the industrialists who supplied the manufactured metals for their success. Joseph Brandon, Joshua's father-in-law, was interested in the New York Central and the Harlem Railroads; Benjamin Nathan, Uriah's brother-in-law, as vice-president of the New York Stock Exchange, was interested in all of the lines made famous by the operations of Daniel Drew, Cornelius Vanderbilt, and James Fiske, although Nathan, who was killed under mysterious conditions after the war, was not linked with the dubious aspects of the railroad ring.

The copper mines received closer attention during the war years when once again the feverish demand for metals for military use turned the attention of the Americans to England. Domestic mines such as the Ridge Copper Company, the Minnesota Mining Company, and the Isle Royale Mining Company, in which the Hendrickses were particularly interested, were also favorites on the mining exchange. Others that appear in Hendricks Brothers records are the Benton Gold Mining Company and the Central of Colorado, as well as the famous Mariposa, the subject of considerable history, and the New York and Nova Scotia Gold Company. The prospect of Tennessee copper, which did not fulfill its first promise, was further reduced by the war. The sale of shares for the copper and gold mines kept the eastern brokers busy. Joshua was brought into the market by his father and father-in-law, but his manner of trading differed from that of his teachers. Instead of trading on the New York exchange he made his transactions through the Lees of Boston or the Florances of Philadelphia. Except for Benjamin Nathan, the New Yorkers had only a vague idea of the nature of the mining interests of Joshua Hendricks.

Another aspect of the Hendrickses' financial interests was banking, which was identified with their firm for more than a half-century. Harmon Hendricks' association with the Hartford Bank was resumed by his sons and continued until shortly before the Civil War. During the war years Uriah Hendricks introduced the practice of utilizing the firm's reserve of refined copper for collateral with the Mechanics Bank of New York. Three hundred thousand pounds of ingot copper, valued at $100,000, served as security for loans up to that amount. This practice was maintained after the death of Uriah Hendricks, and from year to year the amount of ingot copper was increased by quantities of 100,-000 pounds per year. In 1863, to implement an act of Congress that would provide a uniform national currency, Peter Cooper, John J. Astor, Jr., and Edward D. Morgan—a former governor of New York and friend of Lincoln—responded to the National Bank Act to organize a bank in New York. Uriah Hendricks was among the prominent merchants and investors who participated in the organization of the bank.

Against the background of wartime finance, activity in the metal trade continued without abatement. There was no shortage of work, despite the fact that copper and the allied metals became scarcer and competition sharper. The available supply of nonferrous metals did not always meet military and naval needs. However, it was not unusual for Hendricks Brothers to produce $25,000 worth of vital materials for the United States Ordnance Department in a single month and goods amounting to a similar sum for the Quartermaster's Department as well. There was a cooperative exchange between the New Yorkers and their New England competitors, and the names of Crocker, the New Bedford Copper Company, the Reveres, and Lazelle, Perkins & Company appear in the records as purchasers or suppliers of various metals. All of the New York railroads were obtaining supplies from Hendricks Brothers. Bids for government work became highly competitive, and some manufacturers, to assure for themselves a desirable wartime contract, cut prices or sold copper at cost. To correct these practices and break the unhealthy competition, steps were taken to organize the first association of American copper merchants and manufacturers.

The object of the association was to regulate the price of copper and prevent the sale of copper below cost. One of the manufacturers—it has never been determined who—underbid on a government contract with an offer of copper at cost. Lazelle, Perkins & Company, a less successful competitor of both Hendricks Brothers and Revere, accused Hendricks Brothers of violating the gentlemen's agreement. They also charged them with disregarding their obligation to the government by supplying Revere copper which, they said, had "the reputation of an inferior article." Not content with blasting the reputations of America's two outstanding copper firms, the spokesman for Lazelle, Perkins & Company also added that the Crockers, a very active New England firm, cut the price of copper 5¢ per pound on government contracts.

After reading a copy of the charges, which had been forwarded to him, Uriah Hendricks calmly stuck the correspondence into an envelope of the Croton Aqueduct Company and docketed the outside with the following: "placed in this envelope to cool it." John C. Hoadley, of the New Bedford Copper Company, to whom the original letter had been sent, wrote promptly to Uriah Hendricks to assure him that the statements had no influence whatsoever upon the New England companies and that copies of the letter had been made for the other members of the association to familiarize them with the unwarranted conduct of Lazelle, Perkins & Company. Three days later, after Uriah Hendricks had cooled, he wrote to E. W. Barston, of Lazelle, Perkins & Company:

. . . passing over the gasconade relating to yourselves, the perfidity of "Crockers" and leaving the Revere Copper Co. to settle with you, the question of their reputation as inferior manufacturers, with the exception that we deny your libelous assertion that we know them to be so,—we pass to that part where you implicate us. We trust that our position in the community does not necessitate our being jealous of the small share rightfully belonging to you, and that we furnished for the requisition anything, other than agreed, is a gross calumny, and your assertion to that effect is an unqualified falsehood.

We remain with as much respect as we can have for those who allow their temper to get the better of their discretion.

Hendricks Brothers.

Before the copper men of the North had begun their first price war, General Sherman swept through the South, planned the march to the sea, took Savannah and Charleston, and left Columbia a city in flames and smoke. In Virginia on Palm Sunday of 1865 General Robert E. Lee, surrounded by Union troops, surrendered to Ulysses S. Grant. Within weeks the first communication between the Jews of the war-torn South and those of the North was reestablished. Stimulated by the spirit of the spring Passover season and eager to aid the Jewish residents of the devastated southern cities, a movement was organized by Isaac Leeser in Philadelphia to supply Passover *matzot* for the destitute Jews of Columbia and Savannah. A considerable number of northern Jews were hostile to the southerners and refrained from giving aid. But in New York City Uriah Hendricks responded to the appeal by giving the local committee his full support.

Writing to his wife's cousin Augusta Tobias of Liverpool, Uriah informed her that, "The War or rebellion has been crushed and things are once more taking a positive shape and we look for some activity in the fall. All here seem as prosperous as tho there never was a war. Armies said to have been a million men are now disbanding and being paid off."

The first of the family to reach the South after Lee's surrender was Edmund, who had been sent there to resume business with Hendricks Brothers' former southern customers. Arriving in battered Richmond in the fall of 1865, he observed the Jewish holidays in the synagogue Beth Shalome amid a group of former Confederate Jews, some of whom had not yet taken their oaths of allegiance to the federal government. In this lonely and inhospitable atmosphere it must have been a pleasant relief to receive mail from his father. With the Jewish holidays over, Uriah wrote Edmund, it was time for the company to resume full-scale operations in the North and South. In New Jersey water was scarce, Uriah added, orders were behind schedule, the Navy was in immediate need of 2,000 sheets of yellow sheathing for the ship *Onward,* and the city of New Orleans was greatly in need of copper. Prices of metal were still high and brazier copper showed no decline in use or cost.

After a brief stay in Richmond Edmund moved on to Charleston. Meanwhile his brother Francis had reopened his business with the South under the old firm name of Tobias, Hendricks and Company. Their cousin Isaac resumed the partnership with De Cordova on the same basis as formerly. The South, disrupted by the war and humiliated by its losses, was eager to resume trade. Hard cash or gold helped pave the way with the plantation owners. Although Edmund's southern mission was originally to supply the copper needs of the turpentine manufacturers and the distillers of North Carolina, he found that bales of cotton and quantities of rosin could be bought for cash. These goods were wanted by his brother's firm of Tobias, Hendricks and Company. To expedite their purchase an initial shipment of $8,000 in gold was forwarded to Edmund in Charlotte. He purchased what he could and then moved on to Columbia, where the story of wartime devastation was the subject of a letter home: Railroads were not in operation, hard cash was in demand, and moody planters looked with distrust upon northerners.

Because of the antipathy toward northerners even old friends of the Hendrickses were at first reluctant to do business with Edmund, but as weeks passed by, Joseph Lopez, Joseph Tobias, the brother of Jefferson, and the fire-eating Confederate Major Raphael J. Moses became somewhat friendlier to Edmund. Edmund visited many of the South Carolina plantations, called on Moses in Sumter, was bogged down by rain and inadequate transportation, and longed to be back home.

Of the De Leons of Charleston, his second cousins, he saw nothing. Two of them had gained special rank in the Confederacy. Most of those who were old enough to wear the Gray had done so wholeheartedly, Edwin De Leon, a direct descendant of the colonial Uriah Hendricks, had been the United States Consul-General at Alexandria, Egypt, from 1853 until his resignation at the outbreak of the war. He returned to Richmond to run the blockade and to accept a post which made him one of the chief European agents of the Confederacy and brought him into contact with the Confederate agent John Slidell, a direct descendant of the eighteenth-century soap chandler John Slidell, who had done business with the first Uriah Hendricks.

Letters from New York merely increased Edmund's desire to return home. His sister Constance, in an effort to cheer him, described their young sister Emma's debut, the parties at Olaneta's, and the performances which the younger group enjoyed at Wallach's Theatre. His mother's letters supplied other details of Jewish social life in New York City. There were busy Sundays at the Hendrickses, which were always open houses, visits from Solomon Cohen and the Minises from Savannah, Georgia, the Florances from Philadelphia, who were now part of the family, the ever-present English Tobiases, and the Speyers, an enterprising family of Jews from Germany whose success in the gold exchange enabled them to step out of their own social group to join the older Jewish families of New York.

The reopening of trade and the reestablishment of former friendships with the southerners was profitable to both sides. In most cases feelings about the war gradually subsided, and the sale of cotton, rosin, or gunny cloth to Tobias, Hendricks and Company helped reduce sectional distrust. Edmund must have accomplished more than the purchase of plantation products and old copper, for after his return to New York in January, 1866, he was invited by Edgar McMullen of Charleston to return South and become a planter on Edisto Island. To induce Edmund, McMullen added, "The Brass and Lead at Camden I think will alone justify a journey South." But Edmund had had enough of the South, and politely refused the offer. Besides, there was enough snow in Central Park to make winter sleighing pleasant and inviting.

VIII

When Edmund returned home to stay he was admitted as a full member of the firm. Harmon Washington, the youngest brother, was at an age when he could be introduced to the copper world. Uncle Montague, with the aid of his two nephews, managed many of the affairs of Hendricks Brothers, but the senior member of the firm, Uriah Hendricks, retreated into the background. Many of his responsibilities were taken over by Joshua, his oldest son. The initiation of new members to the firm and the promotion of Joshua coincided with changes in the copper trade.

When the war ended, the boom in shipbuilding was over. The government curtailed its naval work and ordered many of the vessels of the blockading fleet to be sold at public auction. The government also disposed of a residue of 5,000 tons of old and unused copper, an amount sufficient for a year's domestic supply. The copper market was confronted by a number of changes: the government surplus competed sharply with the copper shipped East from the Michigan mines, the shipyards had lost a major customer in the United States Navy, and the rolling mills that depended upon the manufacture of sheathing for wooden ships now had to find other uses for their product. Where shipbuilding continued, the iron hull had begun to replace the wooden one and had a decisive effect upon ship construction. The downward trend in copper manufacture was also the result of other aspects of the transition from wartime to peacetime conditions in the industry. These signs of a new crisis that faced the manufacture of copper had their counterparts in the mining industry.

Wartime expansion of copper mining had made large-scale production possible, but with a reduction in domestic consumption, and competition from foreign copper the Michigan mines were confronted with a serious depression. A plentiful supply of copper on both sides of the Atlantic forced the price downward, but the wealth of the Michigan mines produced an opposite reaction among eastern speculators, who plunged heavily into rash investments. Copper stocks were overissued and investment companies ruthlessly promoted the sale of fraudulent stocks. When investment houses were found guilty of embezzlement, the whole copper trade appeared to be in a state of disorganization. The financial districts of the East had good cause for alarm when this state of affairs was disclosed. In 1867 these changing conditions were felt throughout Wall Street and other areas of the investment, banking, and insurance world.

Concealed within the financial web of New York was a resentment against Jews from Germany who had recently achieved success as clothing manufacturers and retail merchants. A number of them had used their hard-earned status to penetrate the elite financial circle of New York during the unstable years that followed the Civil War. Although it was connected, this resentment toward Jews had no relation to the problems confronting the copper industry, but it did finally link itself to the Hendrickses. Further analysis of the position of the Jewish bankers is still speculative, but it is clear that they were not welcome. Though designated as "Jew, German & Co." and "Joseph and Ishmael" by James K. Medbery, a historian of old New York's financial district, the Jewish brokers were unscathed when their competitors on the Gold Exchange resorted only to name-calling. There is no indication of the reaction of Tobias, Hendricks and Company, also members of the Gold Exchange, to these descriptions. Of the 300 who traded in the crowded Gold Room on New Street, there were no more than 20 Jewish bankers or investment brokers, and of this number there were only seven who traded in gold, hardly enough to attract attention beyond the immediate circle of the Bank of New York, or the vicinity of Broad and New Streets. It is not known whether these sentiments were directed at members of the Hendricks family, but the fact that an anti-Semitic mood was simmering within the financial quarter and the large corporations was soon to become a matter of common knowledge wherever newspapers were published and read.

It was the spread of business failures and an uncommonly large number of fires in the clothing trade that brought up the guns that leveled a devastating barrage against Jewish merchants throughout the entire country. Carefully aimed, it struck its mark; this time the Hendrickses were directly involved. The mood had turned into an organized movement, and for the first time in the United States an influential section of the business world set a tone of anti-Semitism.

In the autumn of 1866 six major fire insurance companies adopted resolutions declaring that they would no longer accept "Jew risks." This move proscribed and restricted Jewish merchants and property holders from purchasing fire insurance from the Aetna Insurance Company of Hartford, Connecticut; the Manhattan, the Hanover, the Niagara, the Germania, the Republic of New York City; and the Phoenix of Brooklyn. It was intended to keep this policy confidential among the companies' agents, and for almost six months this was successfully done. It might have passed by without public notice if one of the confidential circulars had not reached Uriah Hendricks, who apparently made it public. Before long the news was spread across the pages of the American press and, though it naturally provoked resentment among Jews, it found some favor among non-Jews. Jewish merchants and businessmen were branded as incendiaries. The barrage had found its mark.

Who released the confidential circular containing this sweeping and devastating pronouncement is not known. Any number of non-Jewish insurance agents whose offices were established throughout the country might have disclosed the contents of the memorandum to a newspaper editor. That the authors of the circular did not anticipate a reaction favorable to Jews from the American press soon became evident when Uriah Hendricks was privately accused of revealing its contents. Hendricks neither admitted nor denied the charge. The Jewish press spoke of "a highly respected Israelite who had taken the lead in this affair in New York," but never disclosed his name. The general press spoke only of local citizens' groups who clamored against the slander. Furthermore, the president of one of the companies— either the Aetna of Hartford or the Phoenix of Brooklyn—expressed surprise at the stinging rebuke he received from Uriah Hendricks condemning the action of the insurance companies, and in weak retaliation he charged that Hendricks had divulged the information to the public.

At first Uriah Hendricks gently reminded the insurance spokesman that it was he who had offended his fellow Christians with his illiberal conduct. Then he retorted to the outrageous suggestion that the Hendrickses ought to have no concern, because the ruling was not intended for them but was meant for the German Jews. Equally offensive to Uriah Hendricks was the contemptuous observation that since the death of Harmon Hendricks, "there are now no good Hebrews left."

Without departing from the carefully chosen language of a disciplined man, Uriah Hendricks reminded his correspondent that he was quite safe in appraising one long since dead but who in his lifetime was known not to brook such an offense. He also pointed out that Harmon Hendricks' "indignation could have been greater than mine and his resentment more rigorous, and none the less . . . for it had emanated from a source that his open hand and purse had resuscitated from the ashes of insolvency." The Aetna of Hartford, the Hartford of Hartford, and the Phoenix of Brooklyn had once counted Harmon Hendricks as a major stockholder and supporter! Now that was forgotten. As for himself, Uriah Hendricks chose to identify himself with the "proscribed" for, as he wrote, "All who observe *that Faith* were thus maligned and insulted for it was *Faith* and not individuals that are Proscribed under it."

Weeks after this correspondence, which was never made public, open meetings by Jews protesting the conduct of the insurance companies were held throughout the North and South. Eventually the proscription was removed. The Hartford Company disclaimed any knowledge of the action and charged their

New York agent with the responsibility. The fire insurance companies retreated with apologies, embarrassment, and chagrin, but some contemporary observers attributed the retreat more to an extensive loss of business than to a change of heart. The challenge offered to the insurance companies to prove a single case of arson by a Jewish merchant passed by unmet.

The Isle Royale Mining Company of Michigan, proceeding as though nothing were disturbing the financial world of the East, elected Uriah Hendricks one of its directors in the summer of 1867, but Hendricks declined the post, claiming that private affairs occupied too much of his time. Meanwhile the Michigan mining companies pressed the legislature of that state to appeal to Congress for protection that would enable the area to continue its advance without fear of competition from foreign copper. Geologists, engineers, investors, and adventurers refused to be turned aside from the common attempt to exploit the abundant resources of the Lake Superior mines. The production of copper was resumed despite the decline in copper consumption.

The ample stocks of raw and finished copper were a real sign of the dull times. Hendricks Brothers' reserve supply of refined copper had been increased to 500,000 pounds; adjustments in the industry were slow and credit was still wavering. The Hendrickses summarized their reaction to this state of affairs in a bit of humorous doggerel that circulated among the brothers:

> We've copper in sheet, copper in block
> But of copper in ingot not a great stock
> We're dealers in spelter and dealers in brass
> But our business ain't worth a spank on the *ass*
> For nobody pays, but all want to borrow
> And promise to pay for certain tomorrow
> Or a little time after, the worry by Gosh!
> Is making an old man, of our promising Josh.

As the future head of Hendricks Brothers, Joshua had much to be concerned with. The tariff panacea had its effect. Domestic copper was unable to counter the free flow of foreign imports, and the drive for additional protection begun in 1867 was finally successful in 1869, when the duty on copper was increased sixfold. The immediate effect of the tariff was in the reduction of foreign ore. This dealt a severe blow to the smelters in the East and forced the closing of smelting works from Baltimore to Boston. Eastern manufacturers were completely dependent upon the Michigan producers, concerned with the additional transportation costs, and for the first time they witnessed the exportation of copper from the United States to Europe. The shifting of control by eastern investment companies and Michigan mines had a direct effect on the manufacturers, but not until another depression and another price corner took place was this to become perfectly clear. Joshua's responsibility to the firm was further increased when, after the insurance episode, a protracted illness of Uriah Hendricks, which completely prevented him from active participation in the business, forewarned his family of his fate.

Death came to Uriah Hendricks, the senior member of the firm, on March 25, 1869, in his sixty-seventh year. The New York press, in its customarily polite obituaries, lamented the passing of a righteous citizen and an outstanding industrialist. The Jewish press wrote at length about the reserve and kind qualities of Uriah Hendricks. It remarked with wonder about this well-known citizen who so seldom participated in public affairs, who never ventured across the threshold of the exclusive clubs of New York even though these doors were open to him, who gracefully turned down directorships of major corporations and accepted public office only during the war years as an expression of patriotic pride. For Jacques Judah Lyons, Shearith Israel's minister, the death of Hendricks was more than the loss of a congregant—it was the loss of a friend. For the American copper trade it was the passing of the oldest member of the industry. Hendricks died as the industry completed the first major cycle of its history: the manufacture of sheathing for ships had come to an end and industrial capitalism had emerged.

EPILOGUE

I

More than a century after the first Uriah Hendricks became an ironmonger in colonial New York his descendants of the fourth generation had brought the name of Hendricks to a peak of importance in the nonferrous metal trade. Traditional family solidarity had eliminated any dependence upon outsiders to manage their business. If, as Francis Hendricks had done, a member of the family chose another business, it was a voluntary matter. As one generation of brothers succeeded another, Francis dissolved his partnership with Thomas Jefferson Tobias and joined the reorganized copper company. His brothers Edmund and Harmon W. were made equal partners and their uncle Montague retired from the firm. Retaining a business in one family was not unique during the nineteenth century, but few American firms were able to boast such longevity. Common business aims, a succession of males, a continued interest in the same trade, and the ability to work in harmony enabled the Hendricks Brothers to keep the firm in family hands. Out of necessity the Reveres had admitted outsiders to their board and to positions of vital control, and other copper firms merged or were absorbed by larger companies. Family continuity and control of the "House of Hendricks" was a tradition known throughout the American copper trade.

Joshua, the oldest son of the oldest son, succeeded his father as head of the firm. From the end of the Civil War, when he first assumed greater responsibility, to the mid-eighties the fate of Hendricks Brothers lay with Joshua. During this period, which coincided with the lengthiest copper slump in the history of the nineteenth century, the industry relied upon the manufacture of brass hardware, bronze castings, heavy sheeting, tubing for flues, and the drawing of copper wire. Within the same period the usefulness of copper was widely recognized. The proliferation of inventions that were dependent upon copper made possible many of the advances in mechanical devices and in machinery. Among these were castings for agricultural equipment and shiny fittings for the new steam fire engines. It was also during the seventies and eighties that copper became known as the best and least expensive conductor of electricity. The brass base of the electric light bulb, the copper-wired armature of the dynamo, the delicate apparatus of the telephone, had begun to consume tremendous quantities of copper wire and placed copper on a par with pig iron. It was at this juncture that the modern history of copper begins. All of these developments coincided with the period of scientific investigation, new mining techniques, and the discovery of rich copper lands in the far West.

Mining was entering its most significant phase in American history. Since the discovery of gold in California in 1849 precious metals were the main object of western miners. Once the California diggings were exhausted the prospectors spread throughout the territorial West in search of richer fields. Little thought was given to copper by the men who hungered to find gold and silver. When precious metals were no longer found in abundance in one area the search continued elsewhere. New discoveries meant new mining towns, towns which were to become famous overnight and which were to vanish as soon as the diggings became sparse or were exhausted. The search in the American West for gold and silver led in turn to the discovery of rich copper areas, the boom towns became permanent centers of industry and community life.

In the late sixties the discovery of gold at Alder Gulch opened Montana Territory to fresh influxes of miners and prospectors. Millions of dollars in gold were panned from the streams and mined from Montana earth. Gold and silver mining became the territory's major industry. Before there was any sign that the new discovery would become barren a rich strike in copper was made by Michael Hickey in 1875 when he staked the now famous Anaconda claim in Butte. The path was opened to a supply of copper that has not been exhausted to the present day. Several years were to pass before the Anaconda site was successfully mined, but meanwhile others were also attracted to the West, lured by its rich wealth in metals. The new copper enterprises were to be rewarded by billions of dollars in the quarter-century that followed. Silver bonanzas, copper kings, and robber barons were typical of an age that was also fruitful in scientific discovery, rich in adventure, and distinguished for its contributions to industry.

Back East, the benefits of these advances were yet to be seen. Hendricks Brothers decided to vacate their Broad Street office for quarters in Cliff Street. The mill at Soho was renamed the Belleville Copper Rolling Mills after a disastrous fire in 1874 destroyed a considerable amount of machinery as well as the Upper Mill. With the family's characteristic energy the mill was immediately rebuilt and was running at full capacity in the remarkably short time of eight weeks. The new mill was large and imposing. Its beautifully landscaped grounds gave it a mag-

Mushroom station at Belleville, New Jersey, about 1895.

nificent appearance and the seventy-five mill workers were in harmony with the new surroundings. Men employed by Harmon Hendricks and Solomon Isaacs, except in rare instances, remained with the firm for life and frequently were succeeded by their descendants, because of the company's policy of maintaining a satisfied labor force. James Moore, who entered the employ of the Soho Copper Works as a young man, eventually became superintendent of the mill and in 1884, at the age of eighty, was still with the firm. The familiar face that confronted the visitor in the thirties met the eye a half-century later. It was this type of experienced man that made possible the rebuilding of the mill after the fire and gave the rolling mill a fine reputation in Belleville. But the rebuilding of the mill marked the end of a period of expansion for Hendricks Brothers. Joshua introduced a conservative policy to contain the assets of the company, a policy which was enforced by the severity of the depression that began in 1873.

During the six years of depression that followed, the tumbling brokerage houses, the bankrupt railroads, and the reorganization of copper mining companies gave Joshua Hendricks little inducement to expand the business into other areas. Joshua rigidly upheld the family business practice of buying all good lots of old and new copper. He was satisfied to follow the custom introduced by his grandfather Harmon of doing business only from a well-stocked warehouse and never departed from this policy.

There were other considerations that compelled him to continue this practice. The tariff of 1869 made foreign imports costly, and the Michigan mines now pooled their resources to control the cost and distribution of copper. This arrangement made procurement difficult, although surplus copper was sold abroad to maintain the domestic price. Under Joshua's direction Hendricks Brothers sought sources other than the Michigan mines, found them, and thereby were enabled to maintain their independent stand. Those metal merchants and manufacturers like Pope, Cole & Company of Baltimore, who were deprived of South American copper by the high tariff and the decline of the seaboard smelters, were compelled to go elsewhere for a regular supply.

Pope, Cole could have purchased their copper from Hendricks

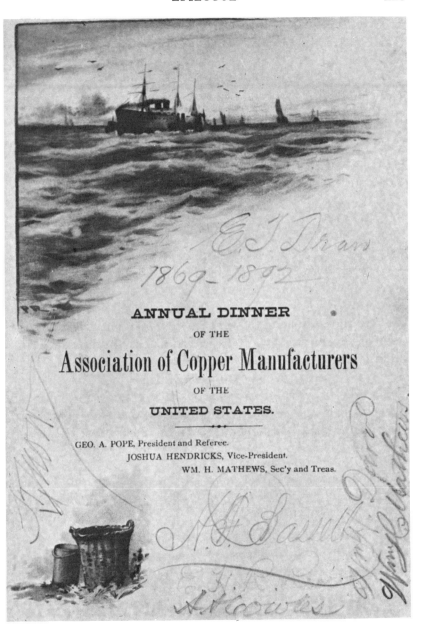

ANNUAL DINNER

OF THE

Association of Copper Manufacturers

OF THE

UNITED STATES.

GEO. A. POPE, President and Referee.
JOSHUA HENDRICKS, Vice-President.
WM. H. MATHEWS, Sec'y and Treas.

Brothers with whom they claimed to have good business relations, but in the mistaken belief that it was the New Yorkers who were attempting to corner the copper market in the East, they resentfully turned elsewhere. As the depression of the seventies continued, Pope, Cole found themselves short of copper and consequently suffered a serious loss of business. Not until they were assured a supply of copper from the Southwest were they able to restore their firm to a sound basis.

Convinced that the New York market was in part controlled by Hendricks Brothers, the owners of the Baltimore mill circulated a letter stating that Hendricks Brothers "with a few other New York operators intend to break down the price of Copper." In a frantic effort to condemn the brothers as hostile to all consumers Pope, Cole wrote to their southwestern source, John R. Magruder of the Mimbres Copper Works, who also sold copper to the New York firm. Magruder, located near Fort Bayard in New Mexico Territory, read about the alleged zeal with which the Hendrickses undersold their competitors and controlled the price of copper. Meanwhile Pope, Cole claimed other New York firms had contracted for all Lake copper scheduled for shipment to the East with the intention of keeping Hendricks Brothers from purchasing Michigan ores. Pressed to obtain sufficient metal to keep their mill in operation Pope, Cole & Company urged Magruder to discontinue selling copper to the house of Hendricks.

. . . all Miners and producers of Copper must be on the same side, as we all want the price maintained. Every ton of Copper Mess. H. Bros. get from outside sources, just so much strengthens their hands in the endeavor to break down the market. All of this we write to explain the present condition of the market. We were not a little surprised to find your Copper there. We do not deny your right to sell where you please; but from what you said when here we expected to get all the Copper you produced, and have paid you what you esteemed full prices at the time of arrival of the different lots,—more than we should have paid had we considered them mere chance lots. If Mess. H. Bros. pay more than we do, we suppose it is only so from the exceptional condition of affairs with them rendering them so exceedingly anxious to get copper from other sources than that one house. Possibly you were not aware the Copper went to them, and that Mess. Reynolds & Griggs may have sent it there without your direction—But we regard it (small though the quantity may be) as none the less injurious to the interests of Copper producers. Please let us know your views on this subject, as also what your prospects are for sending us more Copper.

Magruder forwarded the confidential letter of Pope, Cole to Hendricks Brothers in New York, and if Magruder answered Pope, Cole, a copy of his reply has not been found. Nor has any evidence been discovered to explain why the Baltimoreans did not purchase Lake copper. All of the details of what was clearly a domestic copper corner are not known. But it has been revealed that it was the brokerage firm of Holmes and Lissberger, and not Hendricks Brothers, as Pope, Cole charged, that attempted to purchase the entire annual output of the Lake copper companies, sell it abroad, and guarantee that none of it would be reimported. The brokerage company failed in its attempt and went into bankruptcy. The Baltimore firm had meanwhile overextended itself financially and a few years later also went into bankruptcy. Later their business was acquired by the Baltimore Copper Smelting and Refining Company.

Hendricks Brothers did not allow themselves to be caught between the manipulations of the eastern brokers and the copper pool of the Michigan producers. Their copper collateral was kept in reserve and not placed on the open market. To further guarantee this independent position they obtained an option on the Mimbres Copper Works and eighteen other copper and silver mines owned or operated by Magruder in Grants County, New Mexico. The Association of Copper Manufacturers of Civil War days was no longer effective, and it is doubtful whether it could have counteracted the influence of the banking houses or Lake producers at this time. In an ironical outcome of the conflict between the copper men, George A. Pope and Joshua Hendricks later reorganized the Association of Copper Manufacturers.

Unfortunately the mill fire of 1874 and later losses took with them those accounts, contracts, and correspondence that might have disclosed more of this controversial episode in mining and copper procurement. When the copper corner was broken by an upward swing in business and the American surpluses found a ready domestic market, the flow of copper across the Atlantic was again reversed in 1880. European prices were once again in harmony with American prices. With an uninterrupted supply of copper Hendricks Brothers saw no need to take advantage of the Magruder option in New Mexico, but in turning it down they bypassed the opportunity of acquiring the right to an area that today exceeds all others in the American production of uranium. It was as impossible for Hendricks Brothers to have known of this strategic radioactive material as it was for anyone else concerned with conventional mining. But it was a turning point in the history of Hendricks Brothers when they rejected mining in preference to the manufacturing and importing business.

The upward trend in the copper industry had been caused by railroad construction and locomotive building in the United States and abroad. American interest in foreign railroads had not diminished, and foreign recognition of the techniques and capacities of the Americans to produce sturdy locomotives and fine engines complemented this interest. Since the 1840's, when Austria, France, and Prussia placed their first orders for locomotives in Philadelphia, locomotive building for European countries had increased considerably. At this time Russia appeared with a plan that would enable her to lay rails from the Ukraine to the frozen wasteland in Siberia. Her government contracted for the building of seventy-three locomotives with the New Jersey Locomotive & Machine Company of Paterson, New Jersey. Copper for this undertaking came from the nearby Belleville Copper Rolling Mills. Other European railroad interests involved Baron Maurice de Hirsch, the French-Jewish philanthropist and railroad magnate, who also had extensive investments in European copper, but the nature of this connection has not been determined. Again evidence of the extent of the Hendrickses' locomotive work, like the details of the domestic copper corner, was either destroyed or is today hidden among the records scattered in numerous industrial archives.

Technological improvements and advances in industry had the same effect upon the use of copper in locomotive building as they already had had upon its use in shipbuilding. The amount of copper that went into the manufacture of locomotives was being gradually reduced. The "iron horse," with the exception of its copper flues, remained iron; locomotives and ships had entered the iron age. Copper seemed to be in limbo at the very time that the western mines were on the eve of unprecedented production and when it was about to rise to a new position of prominence in the metal world.

The loss of the ship and locomotive trade was soon compensated for by the gain of another copper market. When Thomas Alva Edison succeeded in developing and completing a system for the distribution of electricity for light and power, copper was the metal that made possible the mechanical aspects of his experiments. Edison's crowning achievement was the Pearl Street generating station, opened in September, 1882, close to the site where Uriah Hendricks had sold copper a century before. The first electric-light power plant in the world illuminated a section of downtown New York the same year that the Anaconda enterprise in Montana was set into motion. Each event was to have a decisive influence on copper in the electrical age. The production of copper wire, which soon became a major factor of the industry, more than replaced the loss of the ship and locomotive trade.

Sensitive to the scientific and technological changes that were affecting mining and refining, smelting and manufacturing, Joshua determined upon a course that would reflect these changes in the firm. A specific area of work was defined for each of his sons. Edgar, the oldest, was admitted to the New York Metal Exchange to master the lively market trading in metals. Henry was sent to the Columbia School of Mines, and when his studies were completed he went on to the study of the chemistry of metallurgy. Clifford set out to familiarize himself with the foreign market as well as the American producers. For the first time in the firm's history its future members obtained a specialized knowledge of independent aspects of the trade before entering the firm.

In 1884 the death of Montague Hendricks, the last of the original firm of Hendricks & Brothers, made necessary another reorganization that would guarantee the future of the company in family hands. Joshua's brothers, Edmund, Francis, and Harmon

W., were bachelors, and Joshua was therefore all the more concerned with his sons' being able to perpetuate the business name of Hendricks Brothers. The stage was being set for a new generation of copper merchants, as copper was being adapted to the uses for which it distinguished itself in the twentieth century.

II

The maturity and stability of the Hendricks family in the financial and industrial world were paralleled by their social attainments. In Uriah Hendricks' lifetime they reached a high-water mark in New York society. Although Uriah had turned away from club life and those socially desirable organizations important to the status of the average businessman, his sons found them acceptable. The New York clubs that came into prominence following the Civil War were already on an exclusive basis. Their common qualifications for membership were, first, an identification inherited with the past of New York City and, second, a business preferably associated with the shipping trade, banking, investments, and railroads. The descendants of shipping merchants were preferred; owners of stately mansions and fine brick houses with other signs of material success were looked upon favorably. Later, religion ranked with channels of trade; Presbyterians came first and the Dutch Reformed and Episcopalians followed. But there were no restrictions preventing Jews from enjoying equal membership when other qualifications were present. The doors were also open to the exclusive turf and yachtsmen's clubs on the same basis. As Jews, the Hendrickses had no need to be concerned about admission on sufferance.

Members of the Hendricks family could therefore be found in the St. Nicholas Society, which required family residence in New York City prior to 1787, or the Union Club, which had similar provisions for membership. Joshua and his brother Edmund were both members of the Union League Club, organized during the Civil War to support the party of Lincoln and the aims of the North. To gratify the family interest in sport, membership in the fashionable Jockey Club enabled Joshua to drive his sulky through Central Park or enjoy the use of his light racing rig. A love for sailing—the Hendrickses had not forgotten their fathers' interest in the sailing craft that went up and down the bordering rivers—made the New York Yacht Club a natural choice. Both Joshua and his son Edgar were members of the Yacht Club for many years, contrary to the belief that there was no Jewish membership in this or similar clubs until recently. Other memberships held by the Hendrickses were in the Vaudeville Club, the Engineers Club and the Fulton Club, of which Joshua was a founder and vice-president.

Many other clubs arose in New York City during the seventies and eighties that attracted a generation of great lawyers, successful merchants, and men of wealth, but despite these club associations the major interest of the Hendricks family went in the direction of helping their fellow men.

An unflagging interest in philanthropy continued to rank high with every branch of a family that provided leadership in communal work and social welfare. Their interest in the consolidation of the Jewish charities, aid to the east European immigrants, and support of the Jews' Hospital—now Mount Sinai—were areas of vital concern to which they turned their energies.

The complexity of philanthropic work resulting from a multiple number of societies that duplicated one another's work led to the merger of a number of agencies under the name of the United Hebrew Charities. One of the five groups to support the merger —while it retained its own independence—was the Hebrew Relief Society. The Hendricks family had been active in the society since its beginning in 1827, and when the United Hebrew Charities was organized in 1874 the same strong support was forthcoming. In the annual reports of the United Hebrew Charities are to be found a record of activity and contribution begun in the seventies that continued far into the twentieth century. Even this new institution found that its resources were quickly drained by the large numbers of needy immigrants from eastern and central Europe.

During the 1880's more than 25,000 Jewish immigrants were arriving in New York annually, and the participation of the Hendricks family in helping this new wave of immigrants was as important as it had been in preceding immigrations. In the past, one generation had befriended the immigrants from England and the West Indies; another cooperated in helping to meet the needs of the Jewish immigrants from Germany; and the generation of the eighties, witnessing the arrival of the east European immigrants, joined with others in relieving their distress.

Refugees from czarist terror were provided with immediate lodging, employment, and sent on to other cities in the belief that the distribution of immigrants was necessary. These pressing conditions called for an organization that could deal directly with the immigrants as soon as they arrived. An ad hoc group had been in existence since September, 1881, but later that year a huge meeting was held to form a society that would invite national support. The result was the organization of the Hebrew Emigrant Aid Societies, with a board consisting of thirty-seven prominent New Yorkers, three of whom were members of the Hendricks family—Isaac Hendricks; his cousin Frederick Nathan; and his brother-in-law Henry S. Henry, who became president of the society.

At Castle Garden, the receiving station for the immigrants, the society provided meals, the services of an employment bureau, proper clothing for those seeking jobs, and transportation funds to bring them to prospective places of employment. The officers of the society worked under difficult conditions—one of them collapsed under the strain—and by doing so revealed their devotion to the immigrants whose habits, language, and customs they did not understand. The subsequent story of the largest Jewish immigration in history is one unto itself.

Support was also given to the newly organized Young Men's Hebrew Association by the copper firm, and the work of individual members of the family was represented in many other organizations. Harmon Hendricks, a son of Montague, was one of the founders of the New York Society for the Prevention of Cruelty to Children and was its vice-president for many years. His sister Sarah—Mrs. Florian Florance—and his two sisters-in-law, Mrs. Albert Hendricks and Mrs. Charles Hendricks, were instrumental in founding the Mount Sinai Training School for Nurses in 1881. Additional support for this program was received from the closely related Gomez, Nathan, and Tobias families.

The traditional family association with Congregation Shearith Israel, which inspired much of the philanthropic work, was strengthened each generation. However the religious conduct of the family was not restricted to only the synagogue and the home. Being away from home and away from the sources of religious comfort and practice in no manner reduced the family's zeal for observance. A typical example is found in the story told by Blanche and Harmon Hendricks while on their honeymoon in Italy in 1877. Arriving in Turin a few days before Passover, they discovered that the only synagogue there was still under construction and, after considerable effort, they located the only shop in the city that sold "passover crackers," the basic ingredient in the observance of that holiday. They were so delighted by this good fortune that they described the experience in a letter back home that told of more than the conventional visits to famous cathedrals or to exhibitions of Renaissance art.

The same devotion and dignity with which the Hendrickses expressed their religious identity characterized their pride in being copper merchants. Edmund, who maintained a close relationship with the Reveres, had begun to gather data on the early years of the Hendricks copper enterprises. In 1889 the Reveres presented Edmund with the correspondence between the first Harmon Hendricks and Paul and Joseph Warren Revere. Edmund was tempted to prepare a history of the firm but never went beyond the first step of gathering together a vast collection of papers that related chiefly to his grandfather, Harmon. Modesty and restraint kept the family from publishing anything about themselves and they permitted only established business facts to be published in the New York trade journals or in the local histories of New Jersey. It was left to the Reveres to publish the first brief account of a family of American copper merchants, which turned out to be a conventional nineteenth-century family panegyric.

By this time Joshua's three sons were completely absorbed with their various responsibilities, which ranged from Michigan timberland, once thought to be a copper region, to Baron de Hirsch's holdings of refined ore, to Rothschild's interest in copper which was now declining. The amazing progress of the whole copper industry was as fascinating to them as it had been to previous generations of the family. The colonial Uriah Hendricks pioneered as an importer of copper; thereafter each generation of sons had to adapt themselves to new business advances and technological changes. Harmon Hendricks, the nineteenth-century pioneer, was identified with ship sheathing; his sons with the production of ship and locomotive work; his grandsons with mining and the electrical age, and the fifth generation with a vast and growing copper empire that concentrated its power in New York City.

The new generation, Joshua's sons Edgar, Clifford, and Henry Harmon Hendricks, faced a New York whose strategic position in the copper import trade was challenged by the increased production of domestic copper. For the first time in its history the firm's hegemony as importers and manufacturers was arrayed against other major firms when they began to operate from New York City. The Crocker Brothers transferred their New England office to the metropolis; the Guggenheims of Philadelphia, who disposed of a business in lace for a western smelting empire, turned to the commercial center of the world; and newcomers who entered the copper brokerage business also gravitated to New York City. Two of the youngest companies, which were rapidly gaining in national prominence, the Montana Ore Purchasing Company and Lewisohn Brothers, recognized the commercial future of electricity and became important holders of Lake copper. Phelps, Dodge & Company, as they entered their sixth decade of business in New York, switched from importing copper to its mining and smelting.

Hendricks Brothers' nationwide reputation, which was inherited by Joshua's sons as importers of ingot copper, block tin, tinners' solder, lead, spelter, and antimony, was on a par with their reputation as manufacturers of copper rivets, wire, locomotive flues, braziers' bolts, and sheathing for all purposes. It was untouched by the new concentration of the copper industry in New York. As a matter of fact, Hendricks Brothers, the second oldest firm in New York—P. Lorillard, the tobacconist, was older by four years—and the oldest in the American copper trade, appeared to be impervious to these shifting conditions.

A plan that might have involved them in smelting and the

subsequent introduction of major changes was rejected by Joshua Hendricks after discussing the idea with William E. Dodge of Phelps, Dodge & Company. At about the same time Phelps, Dodge rejected a plan to build a smelter in the New York area in favor of one that would bring them nearer to the sources of mined copper. It was a turning point in the history of both firms: Phelps, Dodge had committed themselves to mining and smelting, and Joshua Hendricks set aside all such considerations in order to continue the business of importing and manufacturing nonferrous metals. If there was any doubt about this intention in the seventies when the Magruder option was turned down, it was clear in the nineties when Joshua decided not to take positive action. Instead he turned his energies to the reorganization of the Association of Copper Manufacturers, helped draft its constitution, and became its vice-president.

In the belief that an extended trip abroad would improve his health, which had been failing for the past two years, Joshua visited France and England in 1892. In his absence every aspect of the business was in the competent hands of his sons and brothers. While traveling through Europe he received the news of the birth of Edgar's son, Henry, and soon thereafter he returned to New York, although his health was unimproved. A few months later Joshua Hendricks died. Once again there was a shifting and reorganization of the business, which placed Edmund at the head of the firm.

One hundred and thirty years after the business was founded, Edmund and his two surviving brothers, Francis and Harmon W., with their nephews, continued to follow the tradition established by Uriah Hendricks as an importer and by Harmon, who combined industry with the import trade. In addition to their recognized position, the trade attributed to Hendricks Brothers another feature unique in American business history: the descendants of the English houses with which Uriah Hendricks had first begun to trade in the late 1760's were still exporting metals to Hendricks Brothers on the eve of the twentieth century.

The six men, however, were too occupied with the rapid advance of industry to enjoy the armchair laurels of past accomplishments. In the first difficult years of the 1890's Edgar, Clifford, and Henry Harmon experienced the same taste of economic depression as their father and uncles had before them. But once again the firm withstood the battering of hard times.

III

The vitality and enterprise that guaranteed a great promise for the twentieth century were removed by the premature deaths of Joshua Hendricks' three sons. In 1895 the unexpected death of thirty-eight-year-old Edgar Hendricks came as a severe shock to the family and to the firm. Clifford died as unexpectedly a few years later at the age of thirty-nine, and in 1904 Henry Harmon Hendricks died of a heart attack while waiting for the Christopher Street ferry; he was not yet forty-five. With the death of Edgar, the oldest son of the oldest son, the tradition of generations was broken. The effect of the three deaths on the business was not to be felt for several years, but the bachelor uncles who survived their nephews were fully aware of the implications.

Edmund, the senior member of Hendricks Brothers, was fast approaching the Sabbath of his years, but he continued in control of a firm that was producing hundreds of tons of manufactured copper each month. Francis shared Edmund's deep concern for the future of the copper business, but the burden of responsibility was left more to Harmon W., the youngest of the uncles. This was in 1905, when Harmon was approaching the age of sixty and another major copper boom, resulting from the Russo-Japanese War, benefited the American metal trade. The boom was followed by a sharp decline and, before copper returned to a normal level of sales and production, Hendricks Brothers lost through the death of Edmund the benefit of his forty-five years of rich experience in the metal trade. The gentle personality that guided the firm for sixteen years was no longer present to disarm the shrewd purchasing agents who entered the Cliff Street office.

Several years of depressed conditions followed Edmund's death, but they were not at all reminiscent of the stinging crises that had hurt the copper trade in the closing decades of the nineteenth century. The complete absence of business records for this period removes the possibility of any examination of the extent of the mill's activity. Only a gray sadness appears in the tempered mood of the family resulting from the death of Francis Hendricks in 1912. The old generation of Hendricks Brothers had been reduced to one.

Harmon Washington Hendricks, the sole survivor of the firm, characteristically quiet, withdrew to himself. After the death of Francis he was to be seen more often in the Cliff Street office than amid the Victorian furnishings of the Soho estate, and he depended more on trusted employees for the responsibility of the mill's operation than ever before. If he gave consideration to the thought of inviting outsiders to join in the management there is no suggestion of it. The forces of family tradition that tended to control the firm by its own members may have strongly influenced Harmon Washington's conduct. No known attempt was made to reorganize the business or bring to it the necessary spirit conducive to early twentieth-century expansion. While other copper firms looked to mining and smelting and the sources of ore, the surviving Hendricks brother rigidly confined himself to the accepted patterns of nineteenth-century importation, production, and distribution. As the fate of Hendricks Brothers lay with Joshua in the last decades of the preceding century, its course, if there was one, was determined by Harmon Washington on the eve of World War I. All that is clearly known of the firm's management is the single-handed control of the last active descendant. Harmon Washington Hendricks faithfully fulfilled the commitments of the firm but he enjoyed other interests with which his name will be permanently associated.

Harmon Washington's continued residence in New York City —he was living at 270 Park Avenue—was one of semi-retirement. With the exception of his abiding interest in the Museum of the American Indian, of which he was a founding trustee with George G. Heye in 1916, other activity gradually waned. His generous support and his personal aid and encouragement contributed to the success of the new institution. His academic involvement with men eminent in the field of American Indian scholarship and his friendship with Frederick Webb Hodge stirred his imagination far beyond interest in the Belleville mill. The Hawikuh expedition for the study of Pueblo prehistory, a major archeological undertaking, was enthusiastically supported by Harmon Washington and is known as the Hendricks-Hodge Expedition of 1917–1923.

Although Harmon Washington Hendricks was the last of the family to continue in the copper trade, his many relatives were still associated with the civic and philanthropic work that identified them since colonial times. Far into the twentieth century the Hendricks name was prominent in such institutions as the Jewish Theological Seminary of America, the American Jewish Historical Society, and the Jewish Family Service Association. Their number, however, dwindled from year to year and their activity was gradually limited to Congregation Shearith Israel.

IV

Had his father, Edgar, lived it is likely that Henry S. Hendricks might have been raised in the tradition of the copper firm as he was taught to uphold the civic, cultural, and religious traditions of the family. But this is only speculative. There is no question about the record of his other activities. After receiving his primary education Henry S. Hendricks entered Williams College and then transferred to Columbia University, the alma mater of three generations of the Hendricks family.

Columbia was followed by a tour through Europe with his mother, which was shortened by the outbreak of World War I. They turned their steps eastward visiting Egypt and Palestine. Henry was impressed by Palestine, lying fallow under Turkish domination, and the attempts that were being made by its Jewish settlers to redeem the land from a millennium of stagnation.

Back in New York, his career seemed certain when he entered Columbia University law school. The decision to study law now superseded all other interests, and if there was a possibility that he would enter the copper firm there is no evidence of it.

In 1916 he married Rosalie Gomez Nathan. The Hendricks-

Nathan connection, with the mutual social and religious background and the collateral relations that stemmed from the colonial Gomez, Seixas, and Nathan families, was again renewed. A year later Henry S. Hendricks received his law degree and was admitted to the bar. His practice was begun in the firm of his father-in-law, Cardozo and Nathan, the same year that his brother-in-law, Judge Edgar J. Nathan, Jr., began his own career.

When the United States entered World War I in 1917 Hendricks enlisted in the Navy, became a strong supporter of President Woodrow Wilson, and declared himself for a strong Democratic majority in Congress. His political views reflected independent thinking, for he had broken family tradition by urging support for the wartime Democratic administration. The war over, Hendricks promptly reenlisted in the Naval Reserve and returned to the offices of Cardozo and Nathan, with whom he was associated until 1926.

Among the first communal endeavors to involve him at homecoming were the problems facing Mediterranean Jewish immigrants. The resumption of immigration from Turkey, the Balkans, and the Levant was part of the aftermath of the European war. The newcomers settled in the lower East Side amid an older colony of Mediterranean Sephardim established since the early part of the twentieth century. Their habits, religious observances, and language kept them apart from the Jewish immigrants from eastern Europe who were hemmed into the same confining quarter. Overwhelmed by the tremendous flow of immigrants from Slavic lands, this smaller stream was neglected, and to it Henry S. Hendricks turned his attention.

At 133 Eldridge Street a settlement house was established that primarily concerned itself with Mediterranean Jews. The influence of the settlement house grew steadily. Although it never reached the importance of some of the larger Jewish settlement houses, it had characteristics of its own that gave it an honorable reputation in the old East Side. The Mediterranean Sephardim also had the benefit of their own synagogue on Eldridge Street, and settlement house and synagogue functioned side by side. Hendricks' devotion, leadership, and ability to provide the financial aid so desperately needed was rewarded by his election as

Henry S. Hendricks, 1892–1959. *Courtesy of Mrs. Rosalie Nathan Hendricks*

president of the congregation Berith Shalom in June, 1920. It was an honor that he would have graciously bypassed if circumstances had permitted. He was later to use this experience and his influence to bring the small downtown congregation into an auxiliary union with Congregation Shearith Israel, the first and oldest Jewish congregation in the United States.

The groundwork was firmly established. When the tragic news reached New York of the devasting fire that swept Smyrna in 1921, impoverishing many of the relatives of the Eldridge Street community, Hendricks poured his energies into the organization of a special emergency relief committee—independent of other

groups that gave aid to European Jewry in those postwar years—
to aid those in Turkey. In these endeavors he received the warmhearted support of the Shearith Israel Sisterhood in which his
mother, Lilian S. Hendricks, was still a dominant influence.

Without losing sight of this vital overseas aid or of the spiritual
needs of the settlement, Hendricks took advantage of an offer
by the New York Police Department to develop playground
facilities at the corner of Essex and Broome Streets. He equipped
it for a recreation center and assumed the responsibility for its
government. Again, he solicited the aid of the Shearith Israel
Sisterhood and of the congregation's Junior League, over which
he now presided, to make the playground a success for the general use of the entire neighborhood.

Work with the settlement and the emergency relief committee
for the Jews of Smyrna helped Hendricks lay the foundation for
the first Sephardic Jewish organization in New York City. Hendricks became its treasurer and in this capacity enabled Congregation Berith Shalom move from the congested lower East Side
to Harlem and then to the Bronx. His tireless devotion to such
causes made him a desirable candidate for the Board of Trustees
of Shearith Israel. At the time of his election Hendricks was one
of the youngest members to hold such a position in the twentieth
century.

His professional career and his time-consuming communal work
did not distract him from an ever-growing interest in the part
Jews played in American history. Henry S. Hendricks' pride in
his heritage was partly expressed by his activity in the American
Jewish Historical Society. Since 1916, when he first became a
member of the society, he had come to realize that the history
of his own family was also a lesson in American history. Other
members of the Hendricks family had been associated with the
society since its founding in 1892 and they constituted the largest
family unit within the society. Generous gifts of rare material
were made by most of them. In 1920 Henry S. Hendricks was
elected to its executive council and subsequently to the office
of treasurer, in which he faithfully served for almost three decades. Two of the presidents of the society were to sharpen his
interest in the work of the organization, Dr. Cyrus Adler, of the

Congregation Shearith Israel, Central Park West, New York City.
Photograph by Charles Kanarian

Jewish Theological Seminary of America and Dr. A. S. W. Rosenbach, the distinguished bibliophile.

Although Hendricks found himself immersed more and more in the affairs of Shearith Israel, he divided his time among other Jewish institutions as well. His philanthropic and cultural interests made him a candidate for the governing bodies of the Jewish Family Service Association, the Jewish Theological Seminary, and to the Advisory Committee to Jewish Students at Columbia University. Each of these was to have the benefit of his administrative ability and his legal counsel. Activity at Shearith Israel, however, was the fulfillment of his personal desires and the enriching of a family tradition. It was a rare individual who could count seven members of his family who served as president of the historic congregation in the years preceding the American Revolution. These were the Gomezes. Subsequently, members of each generation of the Hendricks family were equally as active from the days of the colonial Mill Street synagogue to the present site at Central Park West. In 1927, when Henry S. Hendricks became president of the nation's oldest congregation, it was in the spirit of perpetuating an old family tradition deeply rooted in American soil. Three years later, when he was unanimously reelected to the presidency, he declined the office by reason of his adherence to the principle of rotation in office. The zeal and fidelity with which he continued to serve as a trustee earned for him a reelection in 1934.

Paralleling his work in the congregation was his participation in the Jewish Family Service Association. The history of the

Interior of Congregation Shearith Israel. *Photograph by Charles Kanarian*

Association, originally the Hebrew Relief Society, is identified from its very beginning with the Tobias, Hendricks, and Nathan families. The same account is to be had of his educational interests, which brought him closer to the inner function of the Jewish Theological Seminary. Here too he was to become a trustee, in 1927. For many years he was the secretary to the Advisory Committee to Jewish Students at Columbia University, and later its vice-president working in close association with such friends as Irving Lehman, Arthur Hays Sulzberger, and the distinguished cousin of his wife, Supreme Court Justice Benjamin Nathan Cardozo.

A busy professional and communal career left little time for relaxation. But in the summertime he was able to turn to yachting, a sport enjoyed by his father and grandfather. Unlike either of them he did not become a member of the exclusive Yacht Club, to which they had belonged, because its social outlook had shifted, but preferred membership in the Knickerbocker Club, which raised no bars to Jewish membership.

During this period of intense activity Henry S. Hendricks witnessed the gradual decline of the copper mill. His great-uncle, Harmon W., with a foreboding view of what was to come, announced his intention of transferring ownership of the stone mansion and more than twenty acres of ground at Belleville to the Park Commission of Essex County. It was in tribute to the town of Belleville, where the Hendrickses, the Gomezes, and the Nathans had lived for more than a century, where many were born and as many died, that Harmon W. determined upon this course of action. This highly desirable area (today the site of Hendricks Park), part of a two-hundred-acre tract, rising in landscaped hillocks, shaded by huge, sturdy trees, was under study for a proposed park and parkway. When Hendricks was approached to sell the land he stated his intention of willing it to the community. The Park Commission, delighted by the news of this unexpected gift, asked whether the transfer of the land could be made within his lifetime. Harmon W. required no inducement. On July 24, 1924, the deed to the land was transferred to Essex County.

On the occasion of his eightieth birthday in 1926, Harmon W.

was honored by Congregation Shearith Israel. Less than two years later, on March 31, 1928, Harmon Washington Hendricks, after returning from a brief trip to the West, died suddenly. The Belleville Copper Rolling Mills closed in honor of the last Hendricks active in the copper industry. Twelve employees of the mill whose services averaged over thirty years attended the funeral, where they listened to Dr. David DeSola Pool deliver a final tribute.

An unusual epilogue to the death of Harmon W., which characterized the Hendrickses' civic-mindedness, is the report published in the *New York Herald-Tribune* of December 4, 1928. On that day the National Metal Exchange interrupted its usual trading for fifteen minutes to auction a pig of tin for the firm of Hendricks Brothers, the proceeds to be used for New York City's United Hospital Fund. It was quickly sold for $1,500 and just as quickly reauctioned for $4,000. As a tribute to the memory of Hendricks Brothers it was placed on permanent exhibit in the lounge of the exchange.

Henry S. Hendricks observed these proceedings thoughtfully. Fully aware of the significance of the Hendricks copper heritage, he became deeply intrigued with the knowledge of the family past. Although his own career was far removed from the copper world, his appreciation of the role of the Hendricks family in the development of early American industry grew with each small discovery of the scattered business and family papers.

After the death of Harmon W. the operation of the mill was gradually reduced and in December, 1938, the Belleville Copper Rolling Mills shut its doors for the last time. It was then, with the closing of the mill, that the correspondence, the old ledgers, the account books, and a mass of miscellaneous papers were discovered at the Cliff Street office, the mill, and the family vault. Twentieth-century records had escaped attention and were either lost or destroyed. The closing of the mill brought to an end 175 years of the Hendrickses participation in the copper industry, the lengthiest epoch of one family's association with the American metal industry, and one that was not equaled in length of years by any other firm engaged in copper manufacturing or mining.

Notes

The Hendricks Collection was deposited in the New-York Historical Society in 1965. It consists of many thousands of items and covers the period from colonial times to the early twentieth century. Documents relating to many collateral branches of the family are also in the collection. Sources other than the papers at the New-York Historical Society are indicated in the notes that follow.

LIST OF ABBREVIATIONS

DAB	*Dictionary of American Biography*
HB	Hendricks Brothers
H & B	Hendricks & Brothers
HH	Harmon Hendricks
HHL I	Harmon Hendricks Ledger I
HHL II	Harmon Hendricks Ledger II
HHLB I	Harmon Hendricks Letter Book I, 1787–1795
HHLB II	Harmon Hendricks Letter Book II, 1798–1801
HHLB III	Harmon Hendricks Letter Book III, 1801–1808
HHLB IV	Harmon Hendricks Letter Book IV, 1809–1825
JE	*Jewish Encyclopedia*
M.W.	Maxwell Whiteman
NY. Gaz.	*New York Gazette*
NY. Mer.	*New York Mercury*
NY. Pac.	*New York Packet*
Occ.	*The Occident, and American Jewish Advocate*
PAJHS	*Publications of the American Jewish Historical Society*
Phila. Gaz.	*Philadelphia Gazette*
Royal Gaz.	*Royal Gazette*
SIA	Shearith Israel Archives
UH	Uriah Hendricks
UH CRB	Uriah Hendricks Cash Record Book
UHL	Uriah Hendricks Ledger
UHLB	Uriah Hendricks Letter Book
UH RBaD	Uriah Hendricks Register of Births and Deaths

1. Uriah Hendricks: Colonial Tradesman and Ironmonger

I

The exact date of Uriah Hendricks' arrival in New York City has not been determined. One of his first newspaper advertisements, *NY. Gaz.*, Jan. 12, 1756, states that he is moving from the Old Slip Market to Hanover Square, suggesting that he may have arrived in one of the fall vessels of 1755. Specific information that could pinpoint the date of Aaron Hendricks' arrival in London has not been found. "Lyons Collection" II, *PAJHS* Vol. 27 (New York, 1920), 387, contains the statement, "Aaron Hendricks went to England [from Holland] when Jews were recalled by Cromwell." This was unlikely for one who died in 1771, more than a century later.

NY. Gaz., Jan. 12, 1756. The merchandise offered for sale by Hendricks consisted of dry goods, articles of clothing, jewelry, trinkets, watches, and an assortment of such hardware that was typical of colonial imports. Jacob Franks (1688–1769) who appears frequently in *UHLB*, described below, settled in New York about 1710 and was prominent in colonial mercantile life.

For a description of Goelet & Curtenius "at the sign of the Golden-Key in Hanover Square," see *NY. Mer.*, Nov. 17, 24, 1755. The iron-mongers' goods consisted of "nails of all sorts and sizes, a great variety of locks and hinges, gun locks and Barrels, carpenters, joiners and shoemakers tools, bellows, shovels and tongs, chimney hooks, cloak pins of many sorts, a large variety of brass furniture, locks &c . . ." *NY. Gaz.*, July 25, 1757, reports that John Dies had just "declined the business of Ironmongery." The same newspaper of April 28, 1755, states that Naphtali Hart Myers moved opposite to the Golden-Key and the *NY. Mer.*, Dec. 29, 1755, offers for sale a variety of his merchandise. For Myers' association with the Great Synagogue in London see Cecil Roth, *Anglo-Jewish Letters* . . . (London, 1938), 155–157. *NY. Gaz.*, Aug. 23, Nov. 22, 1756, provides an additional list of goods Hendricks offered for sale.

For the Myers brothers, Asher, Myer, and Joseph, see Jeanette Rosenbaum, *Myer Myers, Goldsmith* (Philadelphia, 1954). Other references are contained in *UHLB*. Hendricks' letter to his father, Nov. 20, 1758, gives the following reason for Joseph Myers' London trip:

"He is a working silversmith by Trade & came to purchase somethings on his way &c. I believe he brings with him about £600 or £700 sterlg. belonging to him and family. His companion is Sol Marache— likewise on trading voyage and intends to Begin to have a little stock for himself."

Lewis Moses Gomez (1660?–1740) is described by David de Sola Pool, *Portraits Etched in Stone* . . . (New York, 1952), 217–223, as the founder "of the largest and most influential Jewish family in the Jewish community of New York during the eighteenth century." Pool, 334–375, contains the lengthiest and most accurate sketch of Gershom Mendez Seixas, his antecedents and his numerous family. In identifying the Hendricks family of London with the Great Synagogue, Roth, 133–134, mistakenly attributes Uriah's letter to his brother Abram to their father. "Lyons Collection" I, *PAJHS* (New York, 1913), 73, describes the religious ritual.

Uriah Hendricks Letter Book covers the period between May 18, 1758, and Oct. 26, 1759. It is the basic source for this and other references to the Hendricks family. All of the letters are found here relating to Aaron Norden, Moses and Jacob Franks, Uriah's father, his uncle Harmon, his brother Abram, his sisters and other English relatives. The domestic correspondence is directed to the sutlers Benjamin Lyon and Gershon Levy with whom Hendricks came to New York. Of the 120 letters, 17 are interspersed with Hebrew and Yiddish. All of the family matter, the commercial transactions, and some of the religious background are derived from this valuable manuscript. The *Letter Book* is a folio and an undetermined number of the first and last leaves are lacking. Sunday afternoons are described in the letter of June 12 and housing conditions in New York in the letter of June 30 [1758]. Both quotations are from the letter to his brother Abram.

One of the business controversies reflected in the New York press appeared in the *NY. Gaz.*, Sept. 6, 1756. Here, Solomon Hays offered a reward to determine who slandered his "Character and Credit." Hays was a dry goods merchant and had been in the New York trade since the mid-1740's. See *NY. Weekly Post-Boy*, Oct. 28, 1745 *et seqq.*, and *NY. Gaz.*, May 1, 1749.

The opening of Hendricks' third shop is announced in the *NY. Gaz.*, May 2, 1757. Hendricks' notice, "At his store in Hanover Square, two Doors from the Printer hereof," appeared in the *NY. Mer.*, May 12, 1760. The printer, Hugh Gaine, was located at the Bible and Crown on Hanover Square, and John Troup's ironmongery was to the right of Hendricks. *UHLB* [Uriah Hendricks to his father], N.Y., June 30, [1758] for the description of salable clothing.

A note on the genealogy of the Hendricks family is useful here. All hitherto published accounts, including the compendium by Malcolm Stern, *Americans of Jewish Descent* (Cincinnati, 1960) have been set aside in preference to the original manuscript sources. Abram Hendricks who appears in the letter book is otherwise unidentified.

UHLB, undated reference to Levy Cohen's attitude which precedes the letter of June 30 [1758]. Hendricks may have confused this quotation or mixed the verses deliberately for he substituted two words from Proverbs 17:1 for two in Ecclesiastes 4:6. The version however is clear, although the contrast between "one handful" and "two handfuls" is lost. *UHLB* [Uriah Hendricks to his brother Abram], June 30 [1758]. Moses Oppenheim of London is otherwise unidentified. Samson (Sampson) Simson (1725–1773) was the president of Congregation Shearith Israel during 1758 when this letter was written. He was a prominent shipping merchant: Pool, *Portraits*, 502. Mordecai is presumably the Mordecai bar Eliakum referred to in the Hebrew and Judeao-German letter of April [?], 1759, and also without name on Aug. 25, 1758.

Manuel Josephson (1729–1796) was a tradesman in New York City before the American Revolution and he settled in Philadelphia during the war. He was active in Jewish circles and civic affairs: Edwin Wolf 2nd and Maxwell Whiteman, *The History of the Jews of Philadelphia from Colonial Times to the Age of Jackson* (Philadelphia, 1957), for extensive references in index. Josephson to Hayman Levy, Fort Edward, Aug. 10, 1757, *MS.*, Clements Library, Ann Arbor, Mich. Quoted in part by Jacob R. Marcus, *Early American Jewry* Vol. I (Philadelphia, 1951), 77. Marcus does not identify Lyon as the same Lyon referred to by him on pages 225 and 235 as a Canadian sutler.

Lyon, described by Uriah for his father on Aug. 2, and 12, 1758, as "poor little Labe who went with me to New York is taking by the French in late skirmish a going from Fort Edward to the kemp at Lake George." Marcus, 232–234, cites the petitions of Levy Solomons, Benjamin Lyons, Gershon Levy, and Chapman Abrahams of Quebec as copartners. The assumption by Marcus, 225, that Hayman Levy was the procurement source for the sutlers, and that the sutlers obtained financial suport from England is not valid in the light of Hendricks' correspondence. *UHLB*, July 17, 1759: The reference to Solomons & Company is probably to Levy and Ezekiel Solomons, two well known sutlers. Asher and Samuel Isaacs, previously unnoted by Jewish historians, are otherwise unidentified. This letter also reveals the credit relations of the sutlers. *UHLB*, Hendricks to Levy Cohen, N.Y. Apr. 3, 1759: Hendricks supplies a similar description of the sutler in a letter to his father Apr. 5, 1759 which reads, "As for Little *Labe* he shews no Merit of himself to pretend he is Rich & Carries on a Great Trade. I believe they do Trade considerably in the Army & gett Money But on the other hand their Expence to follow the Army is Great, they break *Shabbat* [Hebrew—Sabbath] they Eat *trefot* [Hebrew—non kasher] & have no Regard to Religion therefore hope his addresses to a Certain Lady I heard off has proved Void. . . ." Josephson returned to shopkeeping in New York City: *NY. Mer.,* July 2, 1759.

II

Hayman Levy (1721–1789), merchant, fur trader and army purveyor, advertised extensively in the colonial New York press during the French and Indian War. For pertinent references see *NY. Gaz.,* May 9, 16, and Sept. 19, 1757. Max J. Kohler, "Phases of Jewish Life in New York Before 1800," in *PAJHS* Vol. 3 (New York, 1895), 82 and *ibid.,* 81, for Simson's interest in the Sloop *Good-Intent;* also *NY. Mer.,* Dec. 1, 1755.

Some of the more interesting letters relating to privateering are May 15, July 6, 1758, and Aug. 15, Sept. 13, 1759. The events described by Uriah Hendricks are confirmed by newspaper accounts such as the report of Captain Robert Troup at Antigua in *NY. Mer.,* Apr. 16, 1759. *UHLB,* Nov. 14, 1758, refers to French privateers sweeping the American coast. See also letters of Hendricks dated June 30, July 6, 19, Aug. 12 and Sept. 15, 1758. Edgar A. Maclay, *A History of American Privateers* (New York, 1899), 41–42, for additional references to Captain Troup.

Kohler, 82, for Judah Hays' commission of the *Duke of Cumberland* and *UHLB,* June 30 and Sept. 5, 1758. *NY. Mer.,* July 2, 1759. Hendricks and Hays both advertised that their goods arrived in the *Leopard,* Captain Hunter. The shipment is described in a letter of July 18, 1759, as goods received from Moses Franks and that "Capt.ⁿ Hunter saild under West India Convoy." *NY. Gaz.,* Dec. 1, 1755: June 7, 22, 1756, advertises the kind of goods imported by Samuel Judah and those for Hayman Levy are in the issues of May 16, July 4, 1757. In letters to his father Hendricks describes competition and goods required: *UHLB,* May 31, Oct. 2, Nov. 9, 22, 1758. Others to Aaron Norden and to his brother Abram are dated Dec. 15, 19, 1758. On Mar. 25 and May 15, 1759, Troup's adventures at sea are reported. Abraham Sarzedas, a West Indian Jewish merchant, first appears in New York City when he advertised in the *NY. Gaz.,* Dec. 8, 1752. Hendricks' interest in the sugar trade is expressed in a letter to his father, Sept. 13, 1759. Subsequent references to the West India trade are in letters of Aug. 15 and Sept. 13, 1759. His observation on marriage is under date of June 30, no year, and on Oct. 19, 1759, he announced his decision to settle his business affairs in the colony. The *NY. Mer.*, May 12, 1760, published one of his last spring advertisements. In the fall, the *NY. Mer.*, Oct. 6, 13, 1760, reported "said Hendricks intends next Spring for London," and *NY. Mer.*, May 25, 1761, stated that he is selling goods "near the prime cost for ready money or three months credit he intending shortly for England." Bernard Gratz, the forerunner of an important mercantile family in Philadelphia, is described in Wolf-Whiteman, 36, *et seqq.*

III

Hendricks' residence after he returned to New York is noted in the *NY. Gaz.*, Dec. 19, 1763. No other advertisement appeared under his name during this year. Mordecai Gomez' (1688–1750) obituary appeared in the *NY. Gaz.*, Nov. 5, 1750. Pool, *An Old Faith in the New World* (New York, 1955), 160–161, and in Pool, *Portraits*, 272–274, provides some of the marriage details and antecedents of Eve Esther Gomez.

Uriah Hendricks Register of Births and Deaths [*UH RBaD*] recorded in Hebrew in his ledger is the source for the following entries: Richa was born 11 *Heshvan*, 5524, corresponding to Oct. 18, 1763; Rebecca was born *Shemini Atzeret*, 5525, corresponding to Oct. 18, 1764. Matilda is not listed in the Hebrew family register and precise information of her birth and death has not been found. She is mentioned by Pool, *Portraits*, 479.

According to the *NY. Gaz.*, July 25, 1757, John Dies retired from ironmongery. Goelet later acquired the business of John Troup, *NY. Mer.*, May 12, 1760. The partnership of Peter Goelet and Peter Curtenius was dissolved in 1766 but each continued to advertise regularly in James Rivington's *Royal Gazette* from 1777 to 1782. J. Leander Bishop, *A History of American Manufactures . . .* Vol. I (Philadelphia, 1816), 534, describes Curtenius' foundry in New York in 1775. Hendricks' entry into the copper trade was fixed at 1764 by members of the family in the nineteenth century. Harmon Hendricks established this date with some accuracy: *HHLB IV*, Harmon Hendricks to Mather Parkes, N.Y., Oct. 21 (?), 1821, and the *Shipping and Commercial List and New-York Price-Current*, Dec. 22, 1894, 1–5.

Pool, *Portraits*, 273, relates the merchants' proposal for erecting buildings of brick and stone. Hendricks' new residence was reported in the *NY. Gaz.*, May 9, 1765, and *NY. Mer.*, June 9, 1766, "in Broad Street, next door but one to the corner of Bayard Street. . . ." Naphtali Hart Myers was still in New York according to the *NY Gaz.*, July 31, 1760. Roth, 155–157, shortly thereafter tells of Myers' activity in the Great Synagogue in London. For Hayman Levy's business reverses see Aaron Lopez to Hayman Levy, Newport, Feb. 28, 1765, *Lopez MSS.*, Newport Historical Society. Josephson's activity in Congregation Shearith Israel is better known at this time than his business activity.

UH RBaD: Jocabed Sarah or Sally was born 6 *Heshvan,* 5527, corresponding to Oct. 9, 1766, and Hannah was born 12 *Tevet,* 5528, corresponding to Jan. 12, 1768. Mordecai Gomez Hendricks does not appear in the Hebrew family register. His birth is recorded as of Oct. 12, 1769: *PAJHS* Vol. 27 (New York, 1920), 154. Cecil Roth, "A Jewish Voice for Peace in the War of American Independence," *PAJHS* Vol. 31 (New York, 1928), 37. *UH RBaD,* 17 *Adar,* 5531, corresponding to Mar. 3, 1771, for the birth of Menahem (Harmon).

NY. Gaz., May 6, 1770; *PAJHS* Vol. 26 (New York, 1918), 237–238, noting Hendricks' support of the non-importation resolution. Uriah Hendricks' *Pocket Diary*, London, Dec. 12, 1771, of which only two pages survive, refer to the effects of Aaron Hendricks. This invalidates all statements claiming that Aaron Hendricks died in the colony. The date of his death is recorded in *UH RBaD,* 21 *Adar,* 5531, corresponding to Mar. 7, 1771. Aaron, son of Uriah, was born April 20, 1772, according to *PAJHS* Vol. 27, 155. Information about Matilda has not been found. Charlotte was born 9 *Tishri,* 5534, corresponding to Sept. 26, 1773, *UH RBaD,* where Esther is also recorded on 12 *Sivan,* 5535, corresponding to June 10, 1775. Eve Esther Hendricks' obituary was published in the *NY. Gaz.*, June 29, 1775.

The year in which Hendricks returned to Hanover Square has not been determined. The *NY. Journal,* Oct. 29, 1772, places him near the Custom House but six years later the *Royal Gaz.*, Jan. 17, 1778, states that he is "opposite the sign of the Golden-Key, in Hanover Square." Mordecai Hendricks' death is recorded in *PAJHS* Vol. 27, 286. Pool, *Portraits,* 350–351, provides the best account of Seixas' flight from New York. Wolf-Whiteman, 98, discuss the Gomezes in Philadelphia. Most of the literature presumes that Hendricks remained in New York because of Loyalist sympathies, but no evidence has been uncovered to establish this as a fact. Thomas J. Wertenbaker, *Father Knickerbocker Rebels* (New York, 1948), 106–107; Pool, *An Old Faith,* 46, describes the situation of religious edifices under British occupation. Morris U. Schappes, *A Documentary History of the Jews in the United States, 1654–1875* (New York, 1950), 50–52, for the address of loyalty; J. Solis-Cohen Jr., "Barrak Hays: Controversial Loyalist" *PAJHS* Vol. 45 (New York, 1955), 54–57; *Royal Gaz.*, Feb. 7, 1778, for his commercial activities. Gomez' appearance in Philadelphia is described in Wolf-Whiteman, 120, and that of Phillips, 84–85. Wagg's Loyalist stand is discussed by Roth, 40, and the same author's *Essays and Portraits in Anglo-Jewish History* (Philadelphia, 1962), 165–182.

Hendricks' war-time advertisements appeared in the *Royal Gaz.*, Jan. 17, Mar. 21, 1778, and on Nov. 6, 1782, announcing an "English State Lottery." No other references to Hendricks appeared in the Tory press. The account in the *Royal Gaz.*, Sept. 14, 1782, and other previously unnoted events show the role of Barrak Hays (Barak Hayes) in assuming authority over the Synagogue after the departure of other congregants from New York. Pool, *Portraits*, 274, relates the flogging incident.

Lyon Jonas, London furrier, advertised in the *N.Y. Journal*, Jan. 12, 1775. The *Royal Gaz.*, Jan. 3, 1778, and from time to time until May 5, 1781, carries the advertisements of Levy Simons. Others that appeared in the *Gazette* were: Moses Hart of 994 Water near Peck's Slip, Dec. 11, 1779; Samuel Lazarus, Mar. 22, 1780; Benjamin Raphael of 25 Wall Street, Oct. 24, 1781; Joseph Abrahams, tobacconist and distiller on Queen Street, Oct. 24, 1781; Samuel Levy, a tanner, Aug. 14, 1782; Lyon Hart, May 18, 1782. On Nov. 20, 1782, Isaac Levy states that he has had the honor "to perform before the Nobility of Great Britain, the French King" and so forth. Wertenbaker, 219, gives a brief description of Roubalet's where Levy performed in New York.

Pool, *Portraits*, 50, notes the death of Levy Israel; Roth, 49, notes Wagg's return to London and the *Royal Gaz.* carries Hays' announcement of his intention to return to Europe. Lazarus also announced in the *Royal Gaz.*, Mar. 20, 1780, that he intends to go to London. "Lyons Collection I," 141. It appears that Zuntz's responsibility as president of Shearith Israel was assumed when Barrak Hays left New York and this circumstance may have influenced Zuntz to remain until the British evacuated the city. "Lyons Collection I," 143, is the basis for Hendricks' reception by his returning colleagues. Wolf-Whiteman, 123–124, for Seixas return to New York.

2. Harmon Hendricks: The Making of a Merchant

I

Pieschell and Brogden of London, commission merchants, first appear in *UH L*, in 1797. *HHLB II*, to Pieschell & Brogden, N.Y., Jan. 20, 1799, refers to trade with them prior to this time. In 1809 the firm became Pieschell & Schreiber: *HHLB IV*, Nov. 14, 1809. The last entry in this letter book is May 25, 1822, and thereafter the firm name appears as Schreiber, Hoffman & Hulme. Business with this firm was continued by the sons of Harmon Hendricks. Elva Tooker, *Nathan Trotter Philadelphia Merchant,* 1787–1853 (Cambridge, 1955), makes frequent references to their far flung business accounts.

John Freeman & Company of Bristol, metal merchants, entered into business with Uriah Hendricks about 1769. They operated their own rolling mill and were also commission merchants for the metal trade. The Hendricks Collection contains an extensive correspondence with the Freeman Company and provides important background for the study of many aspects of copper refining, smelting, and distribution. Tooker notes that the Freeman Company was also an important supplier for Nathan Trotter. J. R. Harris, *The Copper King, A Biography of Thomas Williams of Llanidan* (Liverpool, 1964), in addition to valuable references indicates the importance of Freeman as a smelter and establishes the firm's role in the emerging copper world.

The Ledgers covering the period between 1784 when trade with England was resumed and 1796 are not in the Hendricks Collection. If they have survived at all their location is not known. From 1797 to the death of Uriah Hendricks, two extant folio ledgers refer to those which preceded and they suggest a large volume of business with major English metal merchants. Between June and Aug. 1797, the height of the shipping season, Hendricks purchased copper to the amount of £1277.7.8. from John Freeman of Bristol; miscellaneous copper purchases from London totaled £1299.11.1 for June, 1797; from Pieschell and Brogden of London for the period between June, 1791 to July 1798, £3005.8–; woolens from Leeds amounted to £400.18– for the same date; London tin for Sept. was invoiced at £344.5– and leather purchases from London for Sept. 1797 to Feb. 1798 amounted to £554.17.6.

Hendricks' contribution to the Newport synagogue is found in "Lyons Collection II," 185. *UH CRB*, July 29 and Oct. 1, 1793, for references to the dispensary and contributions to the relief fund for French refugees. *Will of John Parmyter*, N.Y., Nov. 25, 1784, which states "all that cash that may remain due me, in the hands of Uriah Hendricks, for the many good services done me in my life time, and give him also Everything that belong to me." *UH CRB*, for reference to Uriah Hendricks paying Parmyter's rent. *UHL*, under 1797 for Lopez reference. The Lopez papers at the Newport Historical Society have not been examined for confirming business transactions.

Articles of Agreement between Uriah Hendricks and Rebekkah Lopez, daughter, singlewoman of Late Aaron Lopez, Oct. 30, 1787: *SIA*. Solomon Levy married Rebecca Hendricks prior to 1793, for their first child, Hetty Grace, was born that year: "Lyons Collection II," 286; Jacob Cohen De Leon married Hannah Oct. 4, 1789: *SIA;* Abraham Gomez married Richa prior to 1797 according to the probate papers of Uriah Hendricks, and Benjamin Gomez married Charlotte on Sept. 13, 1797: Pool, *Portraits*, 432. Sally never married and Esther married Naphtali Judah Nov. 11, 1801, after the death of her father.

Josephson's rabbinic opinion was published in "Lyons Collection II," 187–190 and the request of the New York congregation on a united address to Washington appears in the same work, 217–222. Uriah Hendricks presidency of Shearith Israel is noted in Pool, *Portraits*, 264. Gershon Cohen to UH., Charleston, Oct. 7 and Nov. 26, 1791, *MSS*, reveal the discussion on Jews from Palestine and the methods of Charleston fund raising.

The Constitution and Nominations of the Subscribers to the Tontine Coffee-House (New York, 1796), 16, for Harmon Hendricks' subscription and identifying notes. *UH CRB*, Feb. 12, July 16, 1784; Apr. 22, 1795, for references to John Slidell. Vina Delmar, *The Big Family* (New York, 1961), 38, states that "John Slidell was a Jew" and a "member of Shearith Israel came to bury him."

Edwin De Leon (1818?–1891) a grandson of Jacob Cohen De Leon was Special Commissioner to England and France from the Confederacy: Barnett A. Elzas, *The Jews of South Carolina* (Philadelphia, 1905), 227, 273. John Slidell (1793–1871), *DAB* (New York, 1935), 209–211. Jacob Cohen De Leon (1764–1828) was in Philadelphia in 1789–90 where his son Abraham was born July 15, 1790: *Constitution . . . Tontine Coffee-House*, 16. For the Kingston residence see letter of Hannah De Leon to her father, Jan. 24, [1791]. *UH CRB*, Dec. 2, 1793, and June 10, 1794, which places the De Leons in New York. Letters of Hannah De Leon to UH, Kingston April 17, July 28, [1791]. Jacob De Leon to UH, Kingston, May 22, 1791. *HHLB I*, Sept. 20, 1795 tells of De Leon's trip to Jamaica. *HHL I*, Nov. 10, 1795, and Oct. 1, 1796, contain De Leon's accounts from Spanish town. The DeLeon-Hendricks correspondence from Charleston begins early in 1799.

Solomon Levy (1764–1841) was the son of Hayman Levy the colonial merchant: "Lyons Collection II," 168. *HHLB II*, Feb. 15, 1799, for Solomon Levy's country store.

Information about the birth and death of Abraham Gomez has not been found: "Lyons Collection II," 58; Longworth's *New-York Register, and City Directory* (New York, 1805), 259, lists him as a merchant. The Gomez family Bible under date of Dec. 12, 1822, notes his arrival in Bordeaux, France: "Lyons Collection II," 303, and the Hendricks-Gomez file for 1825 contains two letters written by him from Bordeaux.

Benjamin Gomez (1769–1828) was the son of Matthias Gomez and a second cousin to his wife: Pool, *Portraits*, 432. Lee M. Friedman, "America's First Jewish Bookdealer," in *Jewish Pioneers and Patriots* (Philadelphia, 1942), 180–193.

Naphtali Judah (1774–1855) was the son of Samuel Judah: Pool, *Portraits*, 393; *Occ.* XIII (Philadelphia, 1856), 420, for obituary notice. *UH CRB*, Aug. 10, 17, 1795. John Samuel Ezell, *Fortune's Merry Wheel* (Cambridge, Mass., 1960) for early American lotteries and references to Judah.

II

Rodman Gilder, *The Battery* (Boston, 1936), 123–125. Harmon Hendricks' *Letter Books* supply the basic information as to his personal and commercial activity. They are divided as follows: *HHLB I*, 1787–1795; *HHLB II*, 1798–1801; *HHLB III*, 1801–1808; *HHLB IV*, 1809–1825 Nos. *I* and *III* are incomplete. Hendricks' letters to Solomon Flamengo, Isaac Gabay, David P. Mendez, David Pollander, and other Jewish merchants of the West Indies, his island trip and the years that he was abroad are referred to in *HHLB I* and *II*. His *Ledger* is of equal importance for the study of the West Indian carrying trade in which Jews were active participants during the 1790's. It is one of the most rewarding records of an American firm involved in the carrying trade prior to the War of 1812.

Joseph Correa is referred to in the Hannah De Leon correspondence. *HHL I*, July 23, Oct. 23, 1795, and Jan. 20, June 16, 1796, under entry of United States Office of Discount and Deposit; Dec. 29, 1798, and April 16, 1799, for the Bank of New York. *HHL I*, 1799 contains the first reference to the Manhattan Bank and *HHL II*, June 1799 "To cash paid on the first installment of 100 shares" of the Manhattan Bank. The Manhattan Bank was chartered in 1799 and in 1955 merged with the Chase National to become the Chase Manhattan Bank.

The goods imported by Uriah Hendricks after the Revolution are described in the advertisement of the *NY. Packet*, Dec. 6, 1784. *NY. Directory* facsimile ed. (New York, 1786) lists thirteen metal men of whom eight were ironmongers, two were brass founders, and three were copper- or tinsmiths. The *Ledgers* of Uriah Hendricks contain many more who are not listed in the *Directory*. Some examples of the importance of the older Hendricks' customers can be seen by the following Gershom Jones of Providence, R.I., a pewterer; James Ward of Hartford, Conn., a silversmith; Joseph Beach of Newark, mechanic and craftsman; and John Plum of New Brunswick, N.J., who also was engaged in a number of crafts. Samuel and Judah Myers, sons of the silversmith Myer Myers, were both copper craftsmen, a fact previously unnoted.

Jacob Mark[s] to Alexander Hamilton, Dec. 10, 1794, *Ms:* Record Group No. 59, Genl. Records, Dept. of State, Misc. letters. Oct.–Dec., 1794, in National Archives. This Jacob Marks was not Jewish but was the one associated with Nicholas I. [J.] Roosevelt in various copper schemes. Also *UHL*, 1798, for copper transactions with UH.

Joseph Lopez to *UH*, Newport, Nov. 24, 1791, *MS* on the matter of cheese and its proper ingredients. John Bach McMaster, *A History of the People of the United States* . . . Vol. III (New York, 1892), 499, for the rise in American imports.

UHL, 1797–1798, for the Cohens and Oppenheims of London; *HHLB II*, 1798–1801, for further correspondence. The Pollocks were the first to disappear from the commercial scene and by 1801 the Cohens were no longer involved in the carrying trade. Trade with the Oppenheims continued until it was disrupted by the War of 1812. It was resumed on Mar. 6, 1815.

Mordecai Gomez Wagg (1776–1814) was the son of Abraham Wagg and Rachel Gomez. He is the same Wagg whom Roth, *Essays and Portraits*, 312, describes as Marcus Augustus Wagg. *HHLB II*, June 5, 1797, for business correspondence; *HHL I*, for accounts with Wagg whose name is here spelled as Waage. For correspondence with Daniel Crommelin of Amsterdam see *HHLB II*, Jan. 28, Feb. 22, and Apr. 23, 1797. Daniel was the forerunner of an important branch of the family in nineteenth century Philadelphia. The name is spelled Cromelien and Cromline and Daniel is frequently confused with David. Correspondence with David P. Mendez, Jan. 28 and Mar. 1,

1797, provides interesting views on the conditions and activity of Jews in the West India trade, a subject incompletely explored.

David Lopez (1750–1811) *HHL I*, 36, for the first transaction consisting of an "Adventure to Charleston"; 48, for the account with Jacob Levy of North Carolina begun in 1796 and 48, for Isaacs and Levy of Wilmington, N.C.

James Ward (1768–1856) first appears in *UHL*, 1797. Business with Ward was continued by Harmon and an extensive correspondence between the two men has survived. All of the aspects of the copper trade are to be found here. For Ward as a silversmith see Seymour B. Wyler, *The Book of Old Silver* (New York, 1937), 320, and as a pewterer see J. B. Kerfoot, *American Pewter* (Boston and N.Y., 1924), 65. Ward's individual activity as a Connecticut merchant, tradesman, and craftsman warrants further examination.

Background for Sol Marks and Abraham Cohen in Philadelphia is in Wolf-Whiteman, 206 and 441. *HHLB II*, Mar. 27, 1797. Four letters of Cohen to Hendricks for 1798 and two for 1799 are in the collection. Entries throughout *HHL I* tell of a vast business in cigars.

Lottery sales are reflected in *HHL I* Dec. 20, 1798 to Mar. 26, 1799. If the entry for the cost of copper plates for tickets relates to these sales, then Hendricks was printing lottery tickets also. Other entries in the *Ledger* show that he advertised their sale in the *NY. Gaz.*

Correspondence with David P. Mendez reveals much of Hendricks activity in the island trade in addition to his various personal moods. Letters of special interest are *HHLB II*, Dec. 25, 1797; Jan. 30, May 20, Sept. 13, 1798.

The yellow fever in New York did not reach the epidemic state that it did in Philadelphia, nor did its recurence attract the same historical attention. Walter Judah who died of the fever Sept. 15, 1798, was a brother of Naphtali Judah. See note in *Constitution . . . Tontine Coffee-House*, 73, Hendricks Collection copy. Pool, *Portraits*, 271, suggests this casualty list among Jews but it may have been higher. John H. Powell, *Bring Out Your Dead* (Philadelphia, 1949), 114–139, vividly describes the medical dissention and Harmon Hendricks offered his reasons to Mendez in *HHLB II*, Sept. 13, 1798. "Medicines and advice" are recorded in *UH CRB*, Feb. 6, 1798, and Mar. 20, 1799. For his letter to the English Cohens see *HHLB II*, Oct. 1 and 30, 1798, and Oct. 30 for M. & H. Oppenheim. A one-line obituary notice of Uriah Hendricks appeared in *NY. Gaz.*, Oct. 1, 1798. *HHLB II,* Hendricks to Mendez, June 1(?); Nov. 29, 1798, for Hart's conduct; *UH CRB*, Dec. 5, 1798, for Menahem Isaacs. Samuel and Gabriel Cohen were not given the full details until Apr. 23, 1799.

Correspondence with his father's former colleagues appeared as follows: *HHLB II*, Nov. 1, 1798, to John Freeman & Co.; Nov. 15, 1798, to Thomas Holmes; Dec. 6, 1798, to Stratton and Gibson; Dec. 6, 1798, and Jan. 20, 1799, to Pieschell & Brogden. Beside these, others are also in the letter book. Harmon Hendricks' announcement as administrator is in the *NY. Gaz.*, Dec. 16, 1798. *UHL II*, Dec. 27, 1799, notes the administration papers and conditions of the estate. Marriage contract of Harmon Hendricks and Frances, the daughter of Joshua Isaacs, performed the 4th day of the week, the 11th day of *Sivan*, 5560, *MS, SIA*. Views on Judah and the marriage of Hetty is in *HHLB II*, May 8, 1799, and Feb. 24, Sept. 1, 1800; Aug. 20, 24, 1801.

3. *Copper Crosses the Atlantic*

I

The rise of the American copper industry appears to be a neglected phase of metal history. Up to the present time no adequate study has been found that covers mining, manufacturing, procurement, and the distribution of copper prior to 1815. Isaac F. Marcosson, *Copper Heritage* (New York, 1955), devotes less than twenty pages to the period; Robert G. Cleland, *A History of Phelps Dodge, 1834-1950* (New York, 1952), points to the fact that Anson Greene Phelps (1781–1853) was still a Connecticut saddler until he moved to New York in 1812. Elva Tooker, *Nathan Trotter, Philadelphia Merchant, 1787–1853* (Cambridge, Mass., 1955), 47, states specifically that Trotter did not enter the metal import trade until 1815. The relationship of copper, the carrying trade, and the War of 1812 has not been previously related or examined.

It has not been determined precisely when the English introduced copper sheathing for their vessels, but according to Harris, *The Copper King*, 45, the Royal Navy was sheathing its frigates in 1761. The earliest record of an American vessel found in connection with this study is *The America*, "A Large Copper Bottomed Ship:" *NY. Pac.*, July 5, and Aug. 2, 1784. John H. Morrison, *History of the New York Ship Yards* (New York, 1909), 18, states that the *Empress of China* which cleared New York on Feb. 22, 1784, also had her bottom coppered.

The first of the Orders In Council that affected the copper trade reflected the naval conditions of the war with France. It coincided with Harmon Hendricks' entry into the European carrying trade: John

Freeman to HH, Bristol, Feb. 5, 25; Mar. 20 and Aug. 8, 1799. Harris, 127–130, discusses the subject fully, showing also the English attempt to dominate the copper market.

Victor S. Clark, *History of Manufactures in the United States* (Washington, D.C., 1916), 76; Walter H. Weed, "Copper Deposits of New Jersey," in *Annual Report of the State Geologist for 1902* (Trenton, N.J., 1903), 125; William B. Gates, Jr., *Michigan Copper and Boston Dollars* (Cambridge, Mass., 1951), 7. F. W. Taussig, *The Tariff History of the United States* (New York, 1888), 10–11, 16, where this is discussed in general terms without reference to the metal trade. Copper manufacture was one of the items that received little encouragement. H. J. Habakkuk, *American & British Technology in the 19th Century* (Cambridge, 1962), discusses these problems in the period following the War of 1812 but the specific problem of metallurgy is not a part of this fascinating investigation. Joseph Dorfman, *The Economic Mind in American Civilization* Vol. I (New York, 1946), 323–325.

Nicholas I. Roosevelt (1767–1854) inventor, engineer, and a pioneer in steam navigation, had strong interests in copper mining but in *DAB* Vol. 16 (New York, 1935), 133–134, this activity is not described as fully as in *Appletons' Cyclopaedia of American Biography*, Vol. V (New York, 1888), 317–318. Paul Revere (1735–1818) is portrayed in many biographies and studies but the romantic history by Esther Forbes, *Paul Revere and the World He Lived In* (Boston, 1942) is the most comprehensive, but like Marcosson's book, Revere's contributions to copper manufacture, are glorified, not evaluated.

The Soho Company, Incorporated the 27th November, 1801, by the Assembly of the State of New Jersey (New York, 1802). This valuable fifteen-page pamphlet contains the earliest contemporary description of an American mine and mill, and discusses the technological problems and the economic circumstances confronting Americans at this time. See appendix for the full text. Harris, 37–53, under "The Growth of Anglesey's power" describes the extent of Anglesey's influence in the English industrial world but no mention is made of Soho, Schuyler, or Roosevelt.

This aspect of Revere's copper enterprise has not been previously examined. Its importance cannot be underestimated on account of the fact that it provides new evidence of the Revere operation. Revere the engraver has been carefully studied; the work of his bell foundry properly honored, but the distinction here is between the artist-craftsman and the manufacturer. It is the latter that has suffered neglect in the absence of evidence that would present the problems and difficulties of early American manufacture. The earliest letter in the Hendricks-Revere correspondence is the one written by Hendricks to Revere, N.Y., Feb. 28, 1803, in response to an inquiry from the patriot-engraver. A few of these letters have been published in extract by Mark Bortman in "Paul Revere and Son and their Jewish Correspondents," *PAJHS* Vol. 43 (New York, 1954), 206 *et seqq*. It is a thoroughly unreliable, inaccurate, and distorted interpretation of events: Greenwich Village is placed in Connecticut; spelter is mistaken for saltpeter, ferryboats are confused with steamboats and Harmon Hendricks is thrown into the role of a Revere agent when no such relationship existed. The tone of the article appears to be one that seeks stature for Jews by bringing them into the company of famous early Americans. Harmon Hendricks' position and attitude completely abrogates such an interpretation. See Case XIII, Files 1 to 8 of the Hendricks Collection, for Revere correspondence and documents.

The first ships for which Hendricks supplied copper are reported in *HHL II*. The *Mexicana* on May 10 purchased 958¼ pounds at £1452.0.0; the *Aspasia* on Dec. 24, 1803, purchased 2,835 pounds at £475.5.9. and the *Richmond* on Jan. 23, 1804, purchased 5,878 pounds at £991.18.10. The variation in price suggests copper of different quality.

References to Harmon Hendricks and to Hendricks & Brothers are to be found throughout the records of the Navy Department beginning in 1804 and continuing to 1867. Contracts and correspondence are in the records of the Secretary of the Navy, the Board of Navy Commissioners, and the Bureau of Construction: Reference Service Report, General Services Administration, National Archives and Records Service to Maxwell Whiteman, Sept. 26, 1961. Only a select number of documents from the above record groups have been used for this study.

Revere's queries are noted in the following letters to Harmon Hendricks: Boston, Sept. 27, Oct. 27, and Dec. 26, 1803.

The views of the English agents, exporters and manufacturers expressing the acts of Parliament are discussed in a long series of letters of which the following are essential: Thomas Holmes to Harmon Hendricks, Bristol, Sept. 24, 1799, was the first to describe the difficulties of copper procurement in England. Stratton and Gibson to HH, London, Aug. 7, 1799. John Freeman and Company to HH, Bristol, Feb. 5, "the exportation of all sheathing being prohibited:" Feb. 25, Mar. 12, Mar. 20, and supplement to an earlier letter, Aug. 8,

1799. John Freeman and Company: invoices to HH for plate copper, Mar. 12, 15, Apr. 2, Aug. 4, Sept. 16, 1799, ranging in sums from £400 to £600. "Extracts of Sundry Letters from Liverpool respecting Copper pr Telegraph of Sundry Dates:" Sept. 27 to Oct. 13, 1802, and John Freeman and Company to HH, Bristol, Dec. 24, 1802.

Freeman's summary of accounts with HH in *HHL II*, Sept. 1, 1804, to Nov. 3, 1808, amounted to £566.19.2. Copper purchases from Nov. 1798 to Apr. 1804 amounted to £717.19.10.

The range of Hendricks' imports may be judged also from the large number of commission merchants and manufacturers with whom he did business. In addition to those already mentioned Knox and Hay, of Hull, supplied him with iron and lead; John Hounsfield of Sheffield with rolled iron; William Savill of London with copper; Richard & Thomas Hale of Bristol with pewter and Philip George of Bristol with iron wire.

Harmon Hendricks to Thomas Holmes, N.Y., Apr. 2, 1804, inquiring about fall and spring purchases. John Freeman and Company to HH, Bristol, May 4, 1803, and Bristol Nov. 7, 1804, rejecting Hendricks' proposal to purchase their entire copper production.

HHLB III, Harmon Hendricks to Jacob De Leon, N.Y., Mar. 5, 1804, *HHL II*, April 7, 1806, where Hendricks writes that he "Leased a lot of ground in Bedlow Street and Built a foundery there which cost me 1093.35." Hendricks' New England representatives were David Moses (1776–1858), a son of the prominent New York merchant Isaac Moses, and Joshua Isaacs (1744–1810). Some of the pertinent Moses correspondence is dated Apr. 7, 14, 29, 1806, and in *HHL II*, accounts with Isaacs for the period 1799–1802 are useful. Revere's agreement with Hendricks is dated Boston, Dec. 19, 1805. *HHLB III*, HH to David Moses, N.Y., Mar. 31, Apr. 20, 1806, advising him of the arrangement with Revere.

II

John Atkins to HH, London, Feb. 1, 1802, offering his services as a goldbeater. *HHL II*, 210, for a record of gold leaf importations. Leather shoes of many colors are described in *HHLB II*, HH to Pieschell & Brogden, N.Y., Dec. 6, 1798: also various entries under "Russia Goods." Pieschell & Brogden also carried on trade with the Philadelphia Trotters: See Tooker, *Nathan Trotter*, 43, 77, 80.

Michael and Henry Oppenheim to HH, London, Jan. 21, 1801. In addition to the Oppenheim account, the letter from Knox and Hay, Hull, Nov. 24, 1801, is of equal importance in outlining the changes in European commerce across the seas. All of the correspondence between HH and his English suppliers for the first five years of the nineteenth century provides a broad picture of the variety of imports that reached the United States.

Manuel Judah's correspondence with HH, Richmond, June 10, July 4, 7, Nov. 25 and Dec. 30, 1800, introduces this aspect of the Southern trade. *HHL II*, under "Tobacco from Richmond" and "Tobacco on act. of Manuel Judah Richd." *HHL II*, for the tobacco account with M. & H. Oppenheim of London, Robert Pollock to HH, Richmond, Apr. 4, 1801; Mar. 25, Sept. 26, 1802. Jacob De Leon's cotton and tobacco accounts are listed similarly between 1804 and 1806 as are those of Cohen and Moses of Charleston, 1804, for sales of tobacco and upland cotton. Other references to De Leon in the Southern trade are HH to Jacob De Leon, N.Y., Dec. 31, 1804, and De Leon to HH, Charleston, Aug. 20, 1807. Samuel Grove to HH, Charleston, Aug. 15, 1807, on the slave trade.

Daniel Crommelin & Son to HH, Amsterdam, July 23, 1803, reminding Hendricks that they had not done any business since 1799. When their correspondence was resumed at this time it was mostly for the purchase of metals: *HHLB III*, Jan. 13, 1806, for Dutch-made brass kettles and Oct. 10, for copper and white glass. Bulmer, Horner & Co., to HH, Leeds, July 1, Nov. 2, 1801. Walker, Bulmer & Horner to HH, Leeds, May 30, 1800, advising that the firm has been dissolved and its successors are Bulmer, Horner & Co.

III

Longworth's *American Almanac, New York Register, and City Directory* . . . (New York, 1805), 109, for a list of the brass founders; 126 for copper smiths and 163 for tinners. Bishop, *A History of American Manufactures II*, 586–587, for an account of Robert McQueen's Marine Engine Works. HH to Robert McQueen, N.Y., Aug. 26, 1801, pertaining to legal matters and Longworth, 310, where his first business is listed as the Manhattan Works. *HHL II*, Dec. 9, 1800, for the first appearance of John Youle, and James Ward to HH, Hartford, Oct. 15, 1800, for Ward's preference for the products of Youle's furnace. For other reference to Youle see, Bishop, 587, and Longworth, 417, "proprietor of Air-furnace, Corleaers-hook."

Morrison, *History of the New York Ship Yards*, 21–40, gives sketchy accounts of these shipbuilders. Francis S. Drake, *Dictionary of American Biography* . . . (Boston, 1872), 295, contains an early biography of Henry Eckford. All of these accounts require revision in the light of modern studies. The "Biography of Noah Brown, Ship-Builder" mentioned in Richard C. McKay, *South Street, A Maritime History of New York* (New York, 1934), v, was not available for this study.

Myndert Lancing to HH, Albany, Nov. 17, 1802 and July 14, 1803. Habakkuk, 4–5, discusses this aspect of labor scarcity but for a later period. Henry J. Kauffman, *American Copper & Brass* (Camden, N.J., 1968), 109, for Lancing (Lansing) as coppersmith.

James Ward to HH, Hartford, Oct. 15, 1800. References to tin and the tin trade dominate the extensive Hendricks-Ward correspondence between 1800 and 1812. *HHL II*, for Harford, Partridge & Co. accounts. The same firm was also known as Mark Harford and Brass Wire and Copper Co. It was one of the major ore smelting companies in Bristol, England.

Seth De Wolf to HH, Farmington, Sept. 3, 1803. James Ward to HH, Hartford, Oct. 7, 1805. Luke Reed to Alvan Reed, Sunsbury (?), Sept. 11, 1808: "We have got 30 Pedlers a going out and a part of them have loaded and set out . . . 100 boxes at $18 provided it is good time to send to Baltimore at 4 months and pay interest from date."

The Hendricks-Ward correspondence upon which this is based is in excess of 135 items prior to 1805. During that year the firm name became Ward & Bartholomew. James Ward describes the cleaning of tin for Hendricks, Feb. 20, 1800. Hendricks describes his difficulty with a debtor on July 6, 1801. Myndert Lancing to HH, Albany, Mar. 10, 1803, describes the severity of the previous winter. Other Albany tradesmen with whom Hendricks conducted business were the Fondas and the Spencers. Correspondence with Thomas K. Jones begins in 1801 and continues through 1827.

Hendricks' description of Revere's mill at Canton is in an undated letter to Gurdon S. Mumford, 1807. Mumford, a prominent merchant, was elected to the Ninth Congress for the City of New York in 1805. Between Feb. 28, 1803, and June 1, 1815, fifty-one letters, invoices, and related items of Hendricks and Revere have survived. In addition the Hendricks' letter books, the petitions, and the references by James Ward provide abundant proof that Hendricks was not a mere agent of Revere, or Revere & Son, a thesis projected by Bortman, *op. cit.* Note for example, Paul Revere & Son to HH, Boston, Oct. 27, 1803, where the Reveres agree to purchase 20 cases of soft rolled British sheathing at 43 cents per pound and on four months' credit. Also Paul Revere & Son to HH, Boston, Sept. 6, 1805: "The large bottoms at the price we gave you for the last, 37½ cents, may answer for us, because we can role a large proportion of it into sheets." In the same letter Revere states that they will exchange copper rolled at Canton for English bottoms. Problems resulting from water scarcity were first described by Pieschell & Brogden to HH, London, Oct. 3, 1803, and later noted by Hendricks when he visited the Revere mill at Canton.

A copper engraving of the variety of copper nails was issued in the form of a printed circular by Samuel Guppy of Bristol and sent to HH on Nov. 3, 1806. The printed text is of particular value for its description of the uses of copper nails. No other copy has been located. Depressed business conditions are noted in Paul Revere & Son to HH, Boston, May 26, 1806. Solomon Isaacs (1786–1855), son of Joshua Isaacs and brother-in-law of Harmon Hendricks, writing to HH from New York to Newport, Sept. 12, 1806, speaks of the mosquito.

Competition from Nicholas I. Roosevelt is shown in Paul Revere & Son to HH, Boston, Jan. 30, 1806. Revere's conjecture that it was the Schuyler mine is supported by statements in *DAB, op. cit.* 133–134 and by Talbot Hamlin, *Benjamin Henry Latrobe* (New York, 1955), 169–170. HH to Paul Revere & Son, N.Y., May 31, 1806, advising that the New York market will not compete with the New England market; James Ward to HH, Hartford, Feb. 13, 1809, informing Hendricks that Revere encroached upon the New Yorker's customers.

Frances (Isaacs) Hendricks to HH, N.Y. Dec. 16, 1805, is the earliest in a series of letters to her husband while he was in New England and they reveal some of his experiences. Solomon Isaacs to HH, N.Y. Aug. 26, 29, Sept. 12, 17, 1806, written when Hendricks was in Newport, R. I., with his family.

John McCauley first appears in *HHL II*, May, 1802, and his first letter to Hendricks is dated at Camden, N.J., Oct. 4, 1802. *The Alfred Coxe Prime Directory of Craftsmen*, Vol. 4 (Unpublished) (Philadelphia, 1960), under alphabetical listing has numerous entries for McCauley as a coppersmith. These entries were drawn from the Philadelphia directories from 1791 to 1800. Kauffman, *American Copper and Brass*, 24, also refers to McCauley. Beck and Harvey appear somewhat later in *HHL II*, Dec. 1803 and subsequent correspondence. *Phila. Gaz.* Sept. 18, 1805 *et seqq.* for their frequent advertisements of copper and other nonferrous metals. Abraham Ritter, *Philadelphia and her Merchants* . . . (Philadelphia, 1860), 66–68, for other references to Beck and Harvey. Ritter, 22, 27, for Vanuxem and Clark and the scattered entries in *HHL II*. Solomon Moses (1774–1857) was the son of Isaac Moses of New York and the brother of David previously mentioned. His business activity and social connections are described in Wolf-Whiteman, 347 *et seqq.*

David Gershom Seixas (1788–1864) in *HHLB IV,* July 5, 1811, where he is accepted as a commission merchant for Hendricks and *HHL II,* Aug. 1811 to Apr. 1815. Seixas' unusual career requires further investigation. Beside his work for Hendricks he manufactured crockery between 1812 and 1815, and he was a copper engraver and an inventor. His contributions to science and the mechanical arts are discussed by Isaac Leeser in the *Occ.* Vol. XXII (Philadelphia, 1864), 95–96. Wolf-Whiteman, 331–337, discuss his other activities. Hendricks' *Cash Accounts,* March 2, notes copper purchased from Revere and consigned to Seixas. The *Phila. Gaz.* Aug. 19 and Sept. 16, 1811, carry some of Seixas' copper advertisements.

Solomon Isaacs was established in the commission business when HH wrote to John Freeman on Feb. 9, 1810, "In the Braziery Copper I have intentions of establishing a young man who served his time with me and my kinsman." The Isaacs-Revere correspondence published by Bortman, 217–222, is of course useful provided it is kept apart from the misleading context in which it has been cast. See Case XVI, Files 4 and 5, Hendricks Collection, for Isaacs correspondence prior to 1809.

Some of Hendricks' southern customers are described in *HHLB II,* Sept. 30, 1803. William Dick of Norfolk, Va., was an active account as were John and Theodore Hart of Petersburg, Va., Apr. 1804 to Oct. 1810. Luke Wheeler to HH, Oct. 4, 1805, purchased the bridge copper. Both Jacob Hart, Sr., of New York and Jacob Hart, Jr., of New Orleans are represented in *HHLB II.* The last piece of Hart correspondence is dated July 11, 1809.

IV

In the absence of a reliable work on Jews of the West Indies and their commercial intercourse with the United States the material introduced here like so much for the colonial period, is entirely new. The impact of slave revolts upon the island Jews has received little attention. The decline of the Jewish communities and the problems confronting them are expressed by Aaron Nunes Henriques to HH, Kingston, Jamaica, Apr. 7, May 2, 1802, and David P. Mendez to HH, Kingston, Jam., June 25, 1802. All of the following letters contain information on the interchange between New York, Philadelphia, and the island communities: Benjamin Pereira Da Costa to HH, Kingston,

Jam., Feb. 10, 1802; Aaron N. Henriques to HH, Kingston, Jam., Apr. 7 and May 2, 1802; David P. Mendez to HH, Kingston, Jam., Oct. 17, 1802 and July 19, 1803; Isaac Gabay to HH, Kingston, Jam., Feb. 21, and May 20, 1802. For Isaac Pesoa in Philadelphia see Wolf-Whiteman, 191 and 440, and Pesoa to HH, Philadelphia, April 12 and Nov. 27, 1798. Sixteen other letters of Pesoa are in the Hendricks Collection file for 1799.

HHLB III, HH to Jacob De Leon, N.Y. Dec. 31, 1804, to Abraham G. Wagg of Bristol, Eng., N.Y., Sept. 26, 1803. Although there are many references to Wagg in the United States they must be disregarded in view of this correspondence. Wagg did not come to the United States to fight in the War of 1812 as stated in J. R. Rosenbloom, *A Biographical Dictionary of Early American Jews* (Lexington, 1960), 173.

Although Dr. Samuel Solomon of Liverpool has been described by Wolf-Whiteman, 329, and is referred to occasionally in the Hendricks-Tobias correspondence, his connection with the English copper smelters is only noted by Harris, *The Copper King*, 112. Views on immigrants from England and the islands are expressed in *HHLB II*, HH to Michael and Henry Oppenheim, N.Y., Feb. 5, 1800. *HHLB III*, HH to Aaron N. Henriques, N.Y., May 8, 1803.

HHL II, 400 (family register in French). Hetty was born July 30, 1801; Uriah was born Oct. 8, 1802. Jacob De Leon to HH, Charleston, Dec. 7, 1804, and Oct. 5, 1805, announcing the birth of a daughter. In the above family register the following is recorded: Henry, Born Nov. 28, 1804; Joshua, Mar. 29, 1806, corresponding to *Nissan* 10—the Great Sabbath. Joshua died May 1, 1808. Washington, Nov. 22, 1807; Montague Mordecai, June 12, 1811; Justina Brandela, Nov. 22, 1809, who died May 9, 1823—"She is buried in Oliver Street but no inscription on her white marble tomb." The other daughters were: Frances Henrietta, Nov. 24, 1813, who died May 9, 1817; Emily Grace, Feb. 5, 1816; Hannah, Nov. 18, 1817; Rosalie, Apr. 18, 1820; Selina, Jan. 25, 1823, and Hermione, Apr. 26, 1825. In each entry Hendricks noted the equivalent Hebrew date. Isaac Goldberg, *Major Noah . . .* (Philadelphia, 1936), 17 comments on the Noah children in the Hendricks' household.

Descriptions of home life are contained in *HHLB III*, HH to Jacob De Leon, N.Y., Dec. 31, 1804, and Frances Hendricks to HH, N.Y., Dec. 16, 1805. See also letters of Solomon Isaacs for these years.

Syracuse University holds the papers of the Society for the Reformation of Juvenile Delinquents, the first of its kind in the United States.

Notice of First Regiment, N.Y. Militia, Nov. 15, 1804, and Regimental Orders, N.Y. Militia, 1807, for Hendricks' association. For annual charity accounts, *HHL II*, 103, under "Proffit & Loss." Simeon Levy (1748–1825) and his daughters are described by HH for Jacob De Leon in the letter of Dec. 31, 1804, where his sister Sally is also referred to. Bertram W. Korn, *Benjamin Levy: New Orleans Printer and Publisher* . . . (Portland, Maine, 1961), 12, provides a similar picture of the Levy family. Although Sally is described as insane by Malcolm Stern, *Americans of Jewish Descent,* 79, her conduct, correspondence, mobility, and general demeanor would suggest that by today's interpretations she was eccentric, given to periods of "melancholy," or may have been mentally retarded, but was not insane. In the Dec. 31, 1804, letter to De Leon, cited above, Hendricks advises his brother-in-law that Israel B. Kursheedt is going to Charleston to arbitrate a dispute between Cohen and Moses in order to keep the matter out of court. Some comparison can be made with the Quaker method of arbitration in Tooker, *Trotter,* 60. The Judah episode is spoken of by HH to Michael & Henry Oppenheim of London, Jan. 29, Nov. 4, 1804, and Feb. 13, 1806. *HHLB III*, HH. to Mrs. Abraham Wagg, N.Y. Sept. 13, 1803, to Rachel Wagg, May 25, 1806, and Nov. 1, 1808. "To cost of pulling Samuel Hicks nose," *HHL II*, 198 under "Profit & Loss." For the mercantile position of the Quaker Hicks brothers see Richard C. McKay, *South Street, A Maritime History of New York* (New York, 1934), 32, 45, 46.

V

Numerous versions of the debate over the copper tariff have not taken into consideration the competitive position between importers and would be manufacturers. Bishop, *A History of American Manufactures* II, 126, properly notes the petitions of Revere and the "counter memorials from the merchants, coppersmiths, and braziers of New York and Philadelphia . . . asking a repeal of the duty on spelter, old copper, brass and pewter." F. W. Taussig, *The Tariff History of the United States* (New York, 1888), 10–20, set a standard for future interpreters of the subject. The connection between importation and manufacture was a basic national problem but the beginning of American manufacture cannot with certainty be regarded as a patriotic upsurge as Taussig suggests and others have repeated. To resort to a useful cliché, necessity was the mother of invention, as the Hendricks' papers show. One of the more recent writers on the subject,

Curtis P. Nettels, *The Emergence of a National Economy* (New York, 1962), 274, evaluates briefly the congressional actions of 1808 and 1813. The reasons given by Nettels do not correspond with the views of the copper merchants or show their early efforts in seeking to clarify the tariff and obtain relief from existing duties.

Typical of the numerous documents on marine insurance is John Freeman's explanation to HH, Bristol, Aug. 9, 1805, on the recent captures of neutral ships by French and Spanish cruisers forcing the rise of premiums on American bottoms. Ship manifests for the same period issued by the English merchants reveal difficulties in obtaining space for goods of most descriptions.

The posting of bonds at the Custom House in lieu of paying duties is noted in *HHL I*, Apr. 1807 to Oct. 1812, "for the Decision of the District Court to Determine the legality of the Charge of the Collector for duties on Bottoms & Bolts." Beck & Harvey to HH, Phila. July 11, 1805, ask for information about the designations of copper as raw material. A copy of the petition of Beck & Harvey to Congress, Phila. Feb. 12, 1806, was forwarded to HH. It emphasized the misconstruction of the duty on copper in the laws of 1789 and 1792 and requested the support of HH and other importers. On Feb. 15, HH was further advised that he would not have time to prepare a "Separate Remonstrance to Congress" showing that round bars were considered raw material. Beck & Harvey had requested of HH, Feb. 18, a copy of this petition that could be used by the copper men of Baltimore. John McCauley also advised HH of reaction to the tariff in Philadelphia, and on Sept. 10, 1805, advised that he is cited in the Beck & Harvey address. The document prepared by Hendricks, *Petition of the Importers of Copper in the City of New York. 25th February 1806* is signed by five other merchants: U.S. Naval Records, Record Group No. 233, National Archives. *Debates and Proceedings in the Congress of the United States . . . 9th Congress, 1st Session*, 938. Revere's petitions followed soon after and are reprinted in part by Bortman, 224–225, but without connection to the above cited material. Revere's second petition, Feb. 12, 1807, brought the response from HH which was addressed to Gurdon S. Mumford. Hendricks' second petition, *Memorial, New York Copper Merchants*, was read Nov. 18, 1807. The final report was issued by the Committee of Commerce and Manufactures Jan. 21, 1808, and it is this report upon which most historians have based their opinions on Revere's position.

VI

HH closed his account with Thomas Holmes of Bristol on Mar. 1, 1806. This fact is in itself unimportant but Holmes had taken liberties in acquiring shipping space for others when he was paid to do so for Hendricks. Holmes was also billing Hendricks for services he had not performed and therefore was struck from Hendricks' list of English agents.

The account of the *Ophelia* is based on 376 documents and two small ledgers devoted to her affairs. They are found in document Case VIII of the Hendricks Collection. Additional references to the *Ophelia* are in *HHL I,* for Sept. 16, 1807, and *HHLB III,* Dec. 8, 1806, for letter to Captain Waterman informing him that the restrictions have been lifted and to purchase what he can. HH to John Freeman May 4, July 3 and Sept. 7, 1808, informing him that the *Ophelia* is believed missing for no news of her has reached New York since the ship departed from Bristol. A description of the *Ophelia* is in *HHLB III,* Oct. 31, 1806, recipient unidentified. The Freeman-HH correspondence, 1808, Case IX, File 16, contains a number of references as does all of the miscellaneous correspondence for the period when the *Ophelia* sailed.

VII

According to Timothy Pitkin, *A Statistical View of the Commerce of the United States of America: Its Connection with Agriculture and Manufactures* . . . (Hartford, 1816), 171, 489 ships cleared from Liverpool "and their tonnage one hundred twenty-three thousand five hundred and forty-five." Commercial arrest and the political factors that led to the Embargo are forcefully described by Bradford Perkins, *Prologue to War: England and The United States 1805–1812* (Berkeley, 1961), 140–183 "Embargo: Alternative to War." The reaction to the Embargo is revealed in the following selected correspondence: HH to John Freeman, Apr. 8, Apr. 14, May 4, July 3, Sept. 12, Dec. 6, 1808, and Jan. 4, 1809, to which letter Hendricks added this apology: "Your note reached me after the commencement of Sabbath which prevented my reply until this evening." HH to William Savill & Sons, Apr. 8, 1808, and HH to Philip George, Jr., Nov. 9, 1808. These letters are typical of a correspondence which treats every aspect of the dispute and the difficulties of the carrying trade. John Freeman to HH, Bristol, June 2, 1808, on copper for the New York water works.

How shipping was met under restrictive circumstances is outlined by Philip George to HH, Bristol, Jan. 5, Feb. 7, 21, Aug. 24 and Sept.

9, 1809. Harford, Partridge & Co. to HH, Bristol, Feb. 20, 1809. HH to William Savill & Sons, May 23, 1809. Peter Maze to HH, Bristol, May 1, 1809.

In addition to the standard literature on the Erskine-Smith exchange, Hendricks and his English colleagues expressed the reactions of the merchants to the confusing diplomatic situation: John Freeman to HH, Bristol, June 7, Sept. 16 and 27, 1809. Peter Maze to HH, Bristol, May 1, 1809. William Savill to HH, London, June 7, 1809. HH to William Savill, Apr. 24 and May 23, 1809; to John Freeman, July 31, 1809 and to Pieschell & Brogden, Aug. 1, 1809.

Procurement of tin in England and its distribution in the United States dominate a major portion of Hendricks' trade prior to the War of 1812. He bought and sold tin with zeal. On May 20, 1809, Hendricks advised Peter Maze of Bristol of the quantity he had on hand and placed his advance orders. This followed a prior order on Apr. 8, 1809, of almost 5,000 boxes of tin plates which were to be sent annually on consignment. In seventeen months HH purchased $13,304 of tin and tinplate from Maze. On July 18, 20,000 copper bolt rods of $\frac{5}{8}''$ to $1\frac{1}{4}''$ in diameter were ordered from John Freeman in addition to the 5,000 copper sheets measuring 14" x 48" of various thickness. The seizure of Hendricks' copper was reported to Freeman on July 31, 1809, and Savill's copper which came by way of Puerto Rico was reported Sept. 6, 1809. It was to Pieschell & Schreiber on Nov. 14, 1809, that he wrote of the respect due the American flag while explaining why he no longer pursued the Baltic trade. All of this correspondence is in *HHLB III* and *IV* or in the incoming correspondence to Hendricks for this year.

If Hendricks' statement to Maze, cited in the letter of Apr. 8, 1809, claiming that he had the largest number of customers, was made to induce sales, then his ledger accounts strongly support the claim of the extent of his business. Hendricks wrote similarly to Joseph Walker & Co., Aug. 20, 1810, stating that "in copper Brass Iron tin'd plates and Iron Wire &c. . . . I do larger than any other house in New York." No comprehensive collection of papers on the importation of nonferrous metals or iron specialties, or on the general carrying trade has been located for the period prior to the War of 1812. Until such time that comparable sources are discovered and studied, it would appear that Hendricks' statement must be accepted.

To Peter Maze, Hendricks revealed on Oct. 30, 1809, the manner in which he manipulated a market glutted with tin, and to Harford of

Bristol he explained on May 1, 1810, that with the resumption of trade the tin market may well collapse. Abundant references to the tin trade in the Hendricks Collection offer an opportunity for further study.

VIII

Hendricks' accounting system was inherited with his eighteenth-century training. Records of his investments, his purchases of bank stock and bonds, are scattered throughout his ledgers without obvious system and only occasionally under the heading of the firm or bank from whom he made purchases, or with whom he invested. Some of the stock purchases, investments, and acquisition of shares are to be found in the small *Cash Accounts Book* which begins in 1808. In *HHL I* the following entries are of interest: 102 (1799) and 263 (1799) for the Manhattan Company "to cash paid on the first installment of 100 shares;" 165, 202 and 208 for United States Bank stock and other bank shares; 215 for Louisiana six percent stock. For a typical "Adventure to St. Vincent" see under date of October 27, 1808. Insurance references are also numerous and follow a pattern similar to his other investments such as his subscription in the Mutual Insurance Company of New York for 233 shares on Mar. 17, 1813. In a subsequent summary in *HHL* 382, 393, he lists the Manhattan Fire Insurance Co. and the Phoenix of New York in which he owned 130 and 100 shares respectively. A summary on Feb. 15, 1830, shows that he invested $40,000 in the Hartford Fire Insurance Co.

Money borrowed from Jacob Levy, Jr., *HHL I*, Nov. 18, 1803, to Apr. 31, 1804, amounted to $28,186.92. At the end of 1804 the sum reached $83,763.23 and on 222–226 the figure recorded was $221,-073.77. References to Jacob Levy, whose financial resources are not explained, are frequent prior to the War of 1812.

Some of HH's early real estate acquisitions are listed in *HHL I* as follows: July 16, 1803, 41 lots, Jones & Cornelia St., $1,063.75; Jan. 19, 1804, house at corner Bedford & Commerce, $1,406.16; May 1, 1804, house No. 76 Broad St., $7,131.93; April 7, 1806 "Leased a lot of ground in Bedlow Street and built a foundery there which cost me, $1,093.35;" Sept. (?), 1808, 6 acres of ground on Hudson River, $7,-725.00; Dec. 20, 1813, "Paid Gomperts S. Gomperts atty of Hannah Jacobs 1,000 dolls part consideration for their house No. 65 Greenwich Street bal for $17,500." This also included the house at 61 Greenwich Street. Both were demolished about 1940 when access roads were built for the Battery-Brooklyn Tunnel.

The Hyam M. Salomon correspondence is in the Revere papers at the Massachusetts Historical Society. They are in part reprinted in Bortman, "Paul Revere . . . ," 214–217. *HHL I* under "Merchandise imported in 1815"—"H. M. Salomon, plated metal."

Stedman Adams was introduced by HH to Michael Oppenheim, Oct. 24, 1809. Adams died in London toward the end of 1811: Oppenheim to HH, London, Jan. 15, 1812.

Adam and Noah Brown, ship builders, appear regularly in the papers of HH, see *HHL I,* under 1807. Entries relating to Robert Fulton are in the *Cash Accounts Book* and begin May 12, 1809 and continue through 1813 in the *Ledger.* Throughout 1810 Hendricks' correspondence to his English colleagues show the change brought about by political differences affecting commerce. Business stagnation also contributed to the decline in imports: HH to John Freeman Dec. 2, 1810, and to Peter Maze Dec. 2, 1810, and to Philip George, Dec. 2, 1810. On Aug. 28, and Dec. 8, 1811, HH discusses attempts at metal manufacture in the United States. Gold and silver leaf, a specialty which constantly appears in the records of the Bristol manufacturers and merchants doing business with Hendricks, was according to Bishop, II, 183, introduced in New York in 1812. It was about this time that its import disappeared from the Hendricks *Ledger.* On Jan. 31, 1811, HH wrote to an unidentified correspondent that American business houses were failing, some on account of English failures and others on account of the "Breaking up of the U.S. Bank." The greatest concern about the overseas trade was expressed to John Freeman and Pieschell and Schreiber. On April 14, July 4, and Aug. 7, 28, 1811, HH was requesting Pieschell and Schreiber to invest his funds in negotiable American stocks so that his funds would be available when necessary. The same advice was sent to Freeman on April 14, 1811, and again on Mar. 18, 1812. Both firms were cautioned not to buy U.S. Bank stocks of the southern branches especially those in the states of Georgia and South Carolina and the city of New Orleans. The inability of the two governments to reach an agreement is also shown in the above correspondence. Writing confidentially to Freeman on Feb. 3, 1812, on the matter of Revere's order for 4,000 sheets of copper of various thickness, Hendricks added that it was believed that the American Navy was about to purchase huge quantities of copper. But, he informed Freeman, he had already made huge sales to the Navy whose agents quietly purchased all that they could use. Delayed shipments of copper ordered in England were permitted to enter by congressional action. Orders for sheet iron were placed with Pieschell and Schreiber on Mar. 23, 1812: ". . . not less than 4 feet long or 18 inches wide, the thickness to be from what is known in England by double & treble, or as thick as new Six Penny p^s to the thickness of a new shilling p^s . . ." To P & S, Apr. 5, 1812 HH described the uproar in New York and on Apr. 20, 1812 he expressed to them his inability to comprehend the state of political affairs. It was to them also that he wrote of the declaration of war on July 1, 1812.

4. The Soho Copper Works

I

Harmon Hendricks described his "excursion of a few hundred miles in the interior" for M. Oppenheim on Sept. 14, 1812 and for John Freeman and Pieschell & Schreiber on Sept. 16, 1812. Family background is found in the personal papers cited. The books referred to are in the Hendricks Collection; others are scattered in various collections or are in private hands. Jacob De Leon to HH, Charleston, Oct. 22, 24, 1810, for letters of appreciation of Hendricks' hospitality to Mordecai De Leon. *Catalogue of the Medical Graduates of the University of Pennsylvania* (Philadelphia, 1839) lists Abraham De Leon of South Carolina.

HH reported to John Freeman, Oct. 25, 1812, on the seizure of the *Harriet* and its ultimate release. Solomon I. Isaacs wrote of the "Harriot from Bristol" to Paul Revere & Son, Oct. 1, 1812, describing its fate. The case of the *Independence* was described by HH to Pieschell & Schreiber in letters of Oct. 29 and Nov. 13, 1812, when the privateersmen relinquished their claims. The French adventure which included 38 casks of coffee was under command of Capt. D. G. Gillies, Jan. 14, 1812, and the pig copper was handled by John Higginbottom, July 25, 1813. Five letters were written by Hendricks between Nov. 13, 1812, and Feb. 17, 1813, to his English correspondents; these are crossed through and only one appears to have been sent. No other correspondence is entered in *HHLB IV* until Mar. 6, 1815.

The fluctuation of wartime copper prices was outlined for John Freeman on Dec. 27, 1815, after the war. A study of copper prices in the early decades of the nineteenth century based on the Hendricks records would offer itself as useful background to Orris C. Herfindahl, *Copper Costs and Prices: 1870-1957* (Baltimore, 1959). *Cash Accounts Book*, July 24, 1812, "Subscribed this day to the new 11 million Loan, & took receipt of S. Flewwelling for 16,000$ paid." Although Hendricks' conduct was indicative of the fact that he had no financial concern, he explained his attitude in the following manner to John Freeman, May 1, 1816: "We have been accustomed to large proffitts & ready sale for many succeeding years. We are the more sensible of the change the new state of affairs has brought with them, many of us however may console themselves that if they cannot add to their capital they may live on the interest of it." Hendricks drafted another version of this letter which he did not send.

Some of the banking activity is shown in the following investments which were not limited to his own city and were recorded in his *Cash Accounts Book:* Under July 24, 1812, Hendricks purchased through Ward & Bartholomew $7,950 of shares in the new stock of Hartford, Connecticut's Middletown Bank which was chartered in 1795 and is today the Hartford National Bank and Trust Co. Under the same entry is recorded the purchase of shares in the Pennsylvania Bank (now Company) in Philadelphia. On Dec. 18, 1812, Hendricks withdrew $61,000 from the Manhattan Bank which he used to subscribe to the Hartford Bank, a transaction handled by James Ward. Then, on Mar. 26, 1813, he subscribed $30,000 to the new loan of the Manhattan Bank of which $22,500 was paid by Apr. 1 of that year. On Oct. 13, 1812, the 300 shares of Mechanic Bank stock were acquired for $8,345.80 and on the same date he was negotiating the purchase of shares from the American Bank of New York.

Evidence of Hendricks' bookish interests is spread throughout the collection, frequently in undated entries in his ledgers and in stray notes. Receipts from his brother-in-law, bookseller Benjamin Gomez, and the New York Society Library further document these interests. Scattered throughout the collection are many rare and unrecorded printed prices-current from England and Holland which arrived with the mail sent by his European colleagues. Hendricks subscribed to the various city and trade directories and similar items of mercantile importance which provided useful information to merchants and manufacturers.

The Soho Company, op. cit., and Benjamin Henry Latrobe, *American Copper Mines* (No imprint, 1801), are two valuable contemporary sources that present a picture of the socioeconomic and technological conditions of mining and milling at the beginning of the nineteenth century. Although a history of the Schuyler Mine and its successor, The Soho Company, would add to our knowledge of an intriguing venture, the impressive work by Talbot Hamlin, *Benjamin Henry Latrobe,* 135–136, 169–170, 366–367, supplies the first real understanding of the mine and the mill and the "baffling and fascinating personality" of Nicholas I. Roosevelt. Josiah Hornblower (1720–1809) is another neglected figure in the history of the mine and mill. Hornblower's importance to the mine and mill lay in the machinery and equipment that he brought to colonial Belleville, N.J., assembled illegally, evaded British restrictions, and kept the mine free of water. It was on this site that he erected the first steam engine equipment on American soil. Both *Appletons' Cyclopaedia of American Biography* Vol. III, 264, and *DAB* Vol. IX, 231–232, relied upon William Nelson, *Josiah Hornblower and the First Steam-Engine in America* (Newark, N.J., 1883). The description of Soho is taken from the 1801 pamphlet cited. Fred Graff, *Notes Upon the Water Works of Philadelphia, 1801–1815* (Philadelphia, 1876), 1–2, contains the Cope report on the Soho Works in New Jersey. Hendricks must have been familiar with Roosevelt's financial difficulties, which are not a part of this study, but when he later purchased the Soho Works it was necessary for him to clear Soho of the legal entanglement that involved Roosevelt with the City of Philadelphia. For the Belleville Gunpowder Works see the *Long Island Star*, July 1, 1812.

It is interesting to compare Hendricks' purchase figure of $41,000 with Roosevelt's indebtedness at this time as noted by Hamlin, *Benjamin Henry Latrobe*, 170. Why HH did not use his own name is not clear. One can only speculate upon his desire to keep the copper rolling mill separate from his other business. The *Cash Accounts Book* carries the entries for 20 shares of North River Steam Boat Co. at $2,300, May 20, 1814; 20 shares of York and Jersey Steam Boat Ferry Co. in Hendricks' name, and 20 shares in the name of Solomon I. Isaacs at $4,600, June 3, 1814; and on the same date HH purchased $2,200 in shares of the United Hackensack Bridge Co. Under Jan. 13, 1814, the list of square pigs of copper sent to the mill from Beaver Street is recorded and also the refined copper ready for rolling. His subscription of $42,000 to the $25 million loan, other than $15,000 in his name, was assigned to members of his family.

The wager on the length of the war was made on July 22, 1813, and the death of Major Mordecai Gomez Wagg was reported to HH by Jacob De Leon, Charleston, Oct. 14, 1814. *The Standard of Union*, Feb. 13, 1814 as did other New York papers, carried the notice under "Patriotic Marine Association of New York" that the Browns and others formed a company "for the purpose of annoying the enemies of the United States, and their commerce, by means of private armed vessels."

A waste book according to English definition and usage, is the equivalent of a day book. Entries of all transactions found their way into its pages at the time they were made. Later they were to be carried over to more formal accounting records. Although English in origin, the waste book was popular with American merchants far into the nineteenth century. But most of Hendricks entries were not transferred to his ledger and within its bulky frame is a wellspring of copper history. When used with the ledger and the miscellaneous business records and documents its worth is immediately recognized. The accounts of Adam and Noah Brown appear individually and as a firm appear in the *Waste Book*. On July 25, 1814, the copper purchased by Adam Brown for the *Demologus* is entered. Morrison, *History of the New York Shipyards*, 39 recounts the building of *Fulton the First* or *Demologus*. Demands made upon the shipbuilders who went to Lake Erie are described by John Bach McMaster, *A History of the People of the United States* Vol. IV (New York, 1895), 33.

Mar. 6, 1815, was a day of active correspondence for HH. On that date he addressed himself to Cropper, Benson & Co., Michael Oppenheim, John Freeman and Peter Maze. These letters were followed by others to Pieschell & Schreiber on Mar. 20, 1815, ordering 10 to 20 tons of Peruvian pig copper and reports to the same firm on June 20, 1815, that goods were arriving in abundant quantity without regard to need, particularly exports from Baltic ports. The fear of double duties on imports of Russian goods was expressed to Pieschell & Schreiber on Aug. 8, 1815.

Under "Comparative Wealth of Citizens of New York During and Subsequent to the last War with Great Britain," the tables published by D. T. Valentine, *Manual of the Corporation of the City of New York* (New York, 1864), 755–766, contain useful financial data on the rise of New York's merchant class. Figures on Hendricks do not include the New Jersey mill investments. Foreign goods that arrived unchecked and flooded the American market are discussed in the European correspondence of July 1, 4, and Sept. 30, 1815, and in the summer of 1816 when HH describes the "intemperate importations" for Philip George and John Freeman. S. I. Isaacs & Soho Copper Co. to Paul Revere and Son, June 1, 1815, expresses interest in national achievement in this letter which continues with the statement that conditions "may tempt new manufactures to operate more to our prejudice than any good that might result from a duty of 20 pct."

Valuable background on the growth of the Navy and the shipbuilders' contracts is contained in the *Message from the President of the United States transmitting Reports of the Proceedings which have been had under the "Act for the Gradual Increase of the Navy . . ."* (Washington, 1819). Danish and Swedish copper reached New York in great quantities according to *HHLB IV,* to John Freeman, Mar. 13, 1816.

II

Recent literature devoted to early American technology has not had the benefit of establishing the sources of supply which enabled the forges, furnaces, and founders to operate. The *Waste Book* contains some of the answers for such better known figures as the engineer and inventor, Robert L. Stevens; the stationary engine builder, Robert McQueen of the Marine Engine Works, the brass-founders, James P. Allaire and William Hardenbrook; and the coppersmith John Youle who also manufactured castings. If the ledgers of HH reflect the activities of the carrying trade merchants, the *Waste Book* supplies a multitude of details on the engineers, inventors, and metal men during the first quarter of the nineteenth century. References to many of them are found in the older histories of American industry and technology as well as in the modern work of Carroll W. Pursell, Jr., *Early Stationary Steam Engines in America: A Study in the Migration of a Technology* (Washington, D.C., 1969). Recoppering of the *Demologus* is noted in *HHL II,* July 25, 1814; the *Paragon,* Jan. 2 and June 1, 1815; Mar. 25, 1817 and Mar. 6, 1820; the *Firefly,* Mar. 20, 1815, and Apr. 19, 1821. Adam and Noah Brown's accounts are entered in the *Waste Book,* Sept. 27 and Nov. 5, 1816, and Henry Eckford's, Feb. 8, Apr. 18, and May 1, 1816. What was said of the extent of Hendricks' customers in the general and the tin trade applies to the widespread distribution of copper products other than its marine uses and the sales of such products are to be found throughout the loose miscellaneous orders and ledger entries.

In 1816 George Harley was in business as a coppersmith with a brother. They were located at 95 South Front Street: James Robinson, *Philadelphia Directory for 1816 . . .* (Philadelphia, 1816), unnumbered. The Harleys, as far as correspondence indicates, carried on business with the Hendricks family until the 1850's; Harley papers in the possession of Maxwell Whiteman and *Waste Book,* under 1823. Robert Kid, copper merchant, was located at 101 Walnut Street: Robinson, *supra* and *Waste Book,* 1821. Tooker, *Nathan Trotter,* 93, is unquestionably the only work on the general metal trade after the War of 1812 that offers itself for comparison, although Trotter was not a manufacturer and the slender reference to Hendricks and Soho would hardly suggest Hendricks' far-flung activity.

The sparse accounts of wages paid at Soho are in the *Waste Book:* 1817–$6,244.78; 1819–14 mos. $9,503.70; 1820–15 mos. $5,606.96; Feb. 28, 1822–14 mos. $6,820.47. No further wage data is provided other than "to disbursements here and at Soho for Wood, Mens Wages, cart'ge as may be seen p^r the *Cash Book* from 1 May 1814 to 1 May 1815   7,119.85." The comparison with Revere is based on Forbes, *Paul Revere,* 469, but no precise years are indicated. Later Hendricks noted in the *Waste Book* that between Mar. 1, 1822, and Aug. 1824 "Mens Wages" $15,211.92. On Oct. 16, 1817 he made this computation "Total weight sold since the first sale made at Soho mills for Philadelphia 469,738 lbs. for $175,682.75 total average 37¼ cts. pr. lb." Oliver Evans also appears in the *Waste Book* from Sept. 27 to Nov. 1, 1816. On Oct. 31 is the entry "By 2 pr. rolls $1188.07." See *Indenture* in appendix between James T. Joralemon and Hendricks-Isaacs, July 10, 1817, for nearby mining lands. Samuel Davis, Harmonus Spear to Harmon Hendricks and Solomon I. Isaacs, Lease for mining on the lands of the said Davis and Spear, July 10, 1817.

HHLB IV, to John Holmes, Jr., London, Jan. 17, 1817, Hendricks wrote "I am in want of 12 Ladles made out of wrought ladle moulds to hold from 30 to 40 lbs. of liquid iron, without cracks or flaws ¼ inch thick. Usual size is 9 to 9½ inches diameter, 4½ deep being a half circle. Also 20 lead or reverberating bricks same as last. The ladle moulds you ship'd me all broke in making them into ladles." Hendricks also ordered his bricks from Holmes in the same letter. On Nov. 6, 1817, HH requested of Mather Parkes up to 250 tons of "best house coal to come fresh from the pits."

Typical of Hendricks' inquiry to Freeman on the problems of metallurgy is the following, Oct. 25, 1823: "We are plagued by the qtz of calk or oxide that raises on sheat copper. Our company has often expressed a wish for me to obtain from you the best method to reduce it to metalic copper with the least loss . . ."

Business depression is indicated in correspondence with John Holmes, Jr., Philip George, John Freeman, and Pieschell and Schreiber throughout 1817.

Correspondence with the Navy on this phase of work began Feb. 16, 1816, when Hendricks offered to meet the requirements of the naval advertisement that appeared in the press six days before. On March 20, 1816, he responded to another public advertisement of the Navy, stating that the copper required was available for immediate delivery and on Nov. 27, 1816, he acknowledged John Rodgers' letter detailing naval sheathing needs. There is a considerable amount of unindexed correspondence between Solomon I. Isaacs and the Naval Commissioners, National Archives and Records Service, relating to eleven major contracts during the period between 1815 and 1827. Hendricks letters to John Rodgers describing the expanded mill facilities and the economic problems facing the mill are dated Dec. 1 and Dec. 20, 1816.

A comparative examination of the naval contracts for 1816–1818 is drawn from *Message from the President . . . "Act for the Gradual Increase of the Navy,"* Plate C.

1816, Sept. 16	Levi Hollingsworth	Copper for two 74's and one 44	35 cts. per lb. for all but nails which are 44 cts.
Dec. 30	Paul Revere & Son	do one 74 and two 44's	30 and 32 cts.
	S. J. Isaacs & Soho Copper Co.	do two 74's and one 44	30 and 32 cts.
[?] 17	D. A. Smith	do two 74's and three 44's	29 and 32 cts.
1817, April 5	R. E. Griffith	do two 74's and three 44's	26 cts. and duty on bolts
1818, Sept. 1	Joseph W. Revere	257,750 lbs. of copper bolt rods	36 cts. payable part in old copper at 32 cts.
	Levi Hollingsworth	100,531 lbs. do do	33 cts.
	S. J. Isaacs & Soho Copper Co.	100,631 lbs. do do	33 cts.

The *Waste Book* records further for Sept. 1816 "By copper for govt. 357,087 lbs."

The high price of grain and the decrease in manufacture of new stills was reported by HH to John Freeman, Apr. 27, 1817, and the "languid state of trade" was commented on to Pieschell & Schreiber, Apr. 30, 1817. Mather Parkes was informed that the high cost of English copper was the harbinger of another war. George Dangerfield, *The Era of Good Feelings* (London, 1953), 178 *et seqq.*, gives a picture of the impending panic.

HHLB IV, in a series of letters written in Nov. and Dec. 1817, to Pieschell & Schreiber, Mather Parkes and others, discusses the affairs of the South Americans, the revolution and its effects on copper procurement. Letters for the same period discuss the varying qualities of copper, contrasting European and South American copper. On Jan. 5, 1818, HH wrote to Pieschell & Schreiber telling them of his need for 500 tons annually and on Sept. 10, 1819, the differences between Russian and Swedish copper are mentioned. Copper colors are described in a number of documents: to Pieschell & Schreiber, April 8, 1821, reddish Peruvian and yellowish Mexican are mentioned, and to Newton, Lyon & Co., June 10, 1822, Hendricks advised not to place copper in the lower hold where it changes color as a result of oxidizing. Russian copper, he wrote to Schreiber and Hoffman, Mar. 24, 1825, has a dark plum color and will not sell in the United States. At the same time he wrote to others for such materials that would enable Soho to expand. The new rollers, with Hendricks' specifications, were ordered from John Holmes, Jr. on Dec. 2, 1817. Shipping instructions were sent on May 6, 1818, and on Dec. 8, 1818, the disaster at Barnegat, New Jersey, was reported; then, on Jan. 25, 1819, the rollers were reordered.

Hendricks' summary of the mill's operation in its first three years was recorded in the *Waste Book*, Oct. 16, 1817. Mather Parkes & Co. to whom HH first spoke of his length of service in the copper trade, Feb. 25, 1818, and again on Oct. 20, 1819, was a Liverpool lead, copper and tin merchant. He was a broker who also represented the old English firm of Vivian & Sons after 1819. Business with this firm was begun within a year after the end of the War of 1812 and was continued for many decades. Other references to Mather Parkes are in Tooker, *Nathan Trotter*, 92, 244–245.

Navy accounts are in the *Waste Book*, Jan. 6, 1818; Apr. 10, 11, 1819, *passim*. McKay *South Street*, 128–131, for a brief account of the packet lines. The *Pacific*, believed to be the first of the packets, was built by Isaac Wright & Son and Francis Thompson with copper sheathing supplied by HH. For the regular sailing of the packets see *NY. Gaz. & General Advertiser*, Aug. 5, 1819. Hendricks' own description which included the packets by name, was addressed to

Pieschell & Schreiber, Feb. 10, 1819. *Waste Book*, 1817, "James Monroe, Adam Brown builder, for his own acct. 424 tons." The names of the historic packets appear regularly on hundreds of Hendricks' shipping invoices. Frank O. Braynard, S. S. *Savannah: The Elegant Steam Ship* (Athens, Ga., 1963), 46 published the record of copper purchases for the *Savannah; HHL II*, 1818. Braynard, "Copper for the Savannah of 1818," in *PAJHS* Vol. 48 (1959), 170–176. Although Braynard recognized the tremendous value of the *Waste Book*, he repeats Bortman's misinterpretation of Hendricks. Henry Eckford's colorful career as shipbuilder warrants attention beyond the short accounts in various shipbuilding histories. His flamboyant involvement in the New York political scene, his differences with the Navy Department, and his far flung efforts at establishing navy yards project him beyond the field of local history. Morrison, *History of the New York Shipyards,* 36–37; *Waste Book,* Jan. 3, 1815; May 1, Feb. 8, April 18, 1816, and subsequent references. *Waste Book,* May 13, 1823, for additional copper for the *Robert Fulton.*

James Ward had involved HH in many of the Connecticut banks, insurance companies, and other institutions vital to the commercial interests of that state. One of these was the influential Hartford Bank, established in 1792, of which HH became a director on June 13, 1825. Of the surviving correspondence between Horace Burr and HH which involved him in canal loans, general business loans, and a multitude of financial transactions, the position HH held as a New York representative and director was of particular importance: see for example Hartford Bank files for 1824–25.

Financial difficulties are described in great detail by HH for John Freeman, May 5, 10, 1819, and for Mather Parkes & Co., May 10, 26, and July 10, 1819. Failures among shipping merchants were reported to John Holmes, Jr., June 20, 1819. For his stock negotiations see the letters beginning with Mar. 4, 1819. Real estate acquisition and related items for the same year are in the Miscellaneous Correspondence, Case X, File 16, and should be compared with various ledger entries under *Real Estate.* Valentine, *Manual* . . . 1864, 755–766, for the tax lists and comparative wealth of New York merchants.

Aaron L. Gomez (1801–1865), who later married Hetty Hendricks, entered the employ of HH in 1820. Samuel I. Tobias of the Liverpool family of that name was a son-in-law of the renowned Liverpool quack, Dr. Samuel Solomon, who has been previously mentioned. Tobias was in the watch business in New York by the summer of 1816. On Oct. 6, 1817, he returned to Liverpool with stock certificates and business documents for Hendricks' English colleagues as reported to Mather Parkes & Co. on this date. For David G. Seixas see previous references. Philadelphia's Moss family was prominent in the carrying trade, in the Jewish community and in the civic and political life of the city. For general background see Wolf-Whiteman, 184 *passim*, and for Joseph Lyons Moss, 265–267. A fuller sketch of J. L. Moss, who married into the Hendricks family, is in Henry S. Morais, *The Jews of Philadelphia* (Philadelphia, 1894), 285–286.

HH to John Rodgers, July 23, 1819, appealed for additional naval work. In many of its contracts the Navy, as part of the transaction, exchanged its old or scrap copper. On Feb. 19, 1819, HH notes his proposals for building a steam frigate. Two weeks before he had written to Mather Parkes & Co., July 10, 1819, stating that business failures were widespread and trade reached a standstill, no copper was required. The *NY. Gaz. & General Advertiser*, Sept. 2 and Nov. 9, 1819, also suggests other activity such as the work of the West Point Foundry Association, which had just undertaken steam engine work and machinery castings in iron and brass, and Le Roy, Bayard & Co., who advertised the *Rufus King* "coppered to the bends in Amsterdam with heavy copper," suggesting European competition and Fickett & Crockett's notice of a new, completely coppered brig, on the stocks and ready for launching. Soho products were made available for these American undertakings. Men's wages, extracted from the *Waste Book* for these years were previously noted. Revere purchased pure copper from HH Feb. 1 and Mar. 2, 1819. A description of pickle dust was sent to Mather Parkes & Co. to whom it was offered on Sept. 4, 1823. The cost was six pounds per dollar.

III

The revival of business in the 1820's can be thoroughly documented by hundreds of examples for that year. Christian Bergh, to whom HH supplied copper, appears in *HHL II*, June 10, 1808, at Corlears Hook. Bergh was with the men who went to the Lakes during the War of 1812. Upon his return he resumed business with HH on Dec. 26, 1815. Bergh built the *Don Quixote* in 1823–24, purchasing his copper from HH. Foreman Cheeseman worked as a ship carpenter for the Brown brothers. Their joint accounts are listed in *HHL II*, July 10, 1803–Apr. 2, 1807. He shared an interest in the ship *Empress*, for which HH provided building funds and funds for its first voyage: *HHL II* Mar. 26, 1821, to Mar. 27, 1827. Henry Eckford appears regularly in the *Waste Book*, May 20, 1820; Nov. 27, 1821; Jan. 13 and Feb. 28, 1826, when he acquired copper for the steamboat *Nautilus*. The ledger account contains this relevant statement "May 14, 1829—to cash loaned him for 5 years at 7 pr ct on 6 lots 30 feet each now leased and occupied by George Fordham as a steam saw mill for 15 years @ 500$ per Ann." The *China* was an East Indiaman of 533 tons. Accounts of Adam and Noah Brown are scattered throughout the business records during Hendricks' lifetime. For Fickett and Crockett see in addition to the *Savannah* notes, *Waste Book*, Nov. 16, 1820, and Oct. 31, 1821. Isaac and Sidney Wright became active during the War of 1812, *Waste Book*, Sept. 27, 1815, to May 24, 1821, when they were building for Francis Thompson founder of the Black Ball Packet Line. On the Wrights, HH noted May 1, 1825 "Leased them my ½ acre Lot on Oldport road & 3d Avenue at Kips Bay 54 pr ann & taxes, 50 feet by 400 feet from year to year . . . no papers passed." Isaac Webb and his son William H. entered shipbuilding near Corlears Hook about 1818. The older Webb had worked for Eckford. For one of their accounts see *Waste Book* under 1823. HH to John Freeman, Jan. 20, 1820, notes with pride the speed of American packets.

More and more Hendricks' business connections with the shipbuilders assumed a different relationship. Not only did the Wrights and Eckford become dependent upon him for the expansion of their yards and for personal loans but so did Noah Brown. On Aug. 7, 1817, HH wrote in his Ledger, "to cash loaned him on his Bond & Mortgage pay'e 2 years on his shipyard, 31 lots and his home . . . 16 lots at Corlears Hook @ 7 pr ct half yearly." "Sep. 4, 1819 to do. Loaned him additional cash on the above properties and took 2d mortgage, with 1 house & 2 lotts in North Street." The current rate of interest in New York City at this time was 12%.

News of the tariff debate by Congress was reported to Pieschell & Schreiber, Apr. 5, 1820, and on Apr. 27, 1820, its consideration of a Bankruptcy Act. Power of attorney for Freeman was arranged with James C. Neilson of Baltimore, Jan. 2, 1820. When Congress failed to pass the insolvency act it was reported by HH to John Freeman, June 8, 1820. Hendricks vs. Robinson is represented by 54 documents including account book material in Case X.

Reconsideration of the tariff alarmed the copper men. Hendricks petition, *HHLB IV*, Feb. ?, 1820, was signed by Solomon I. Isaacs and Soho Copper Company. On June 8, 1820, HH informed John Freeman that Congress did not impose a duty on copper sheathing. Henry Meigs to HH, Washington, July 11, 1820, advising that the Senate passed a bill to relieve him and Denton Little of their claim.

For additional background see tariff file, 1811. HH to Robert Swartwout, Mar. 28, 1820, explaining the copper process and its cost.

Governor Aaron Ogden appears in the *Waste Book*, Feb. 19, 1816. Transactions betwen Fulton and Livingston also appear here beginning June 30, 1815. See also the accounts with John R. Livingston, 1816–1827.

Between Jan. 25 and Feb. 6, 1821, the copper for the steamboats *Mexicana, Richmond,* and *Chancellor Livingston* is recorded in the *Waste Book.* James B. Allaire to Soho Copper Works, Feb. ?, 1820, for bolts and copper for a small engine. John J. Gray and Bros. to HH, Mar. 10, 1821, requesting copper which is in great demand in the Kentucky area. For the Wrights' contract for a new ship, see correspondence in Case XI, File 10, where confirmation is also found for Le Roy Bayard's "American Rolled and Refined Copper." Papers relating to the *Empress* are in Case XI, File 12, and Case XII, Files 13–18, provide a comprehensive picture of the year's activity.

Melville's description of coppered ships is in the opening pages of *Redburn.*

A lengthy account of the superior qualities of zinc sheathing and its cost appeared in the *NY. Gaz. & General Advertiser,* Sept. 4, 1819. Naval and other government contracts, purchasing arrangements, and points of delivery are detailed in the *Waste Book,* Jan. 5, 1822, for the *Constitution* at Norfolk; July 21, 1822, for delivery of copper to Washington; Aug. 15 and Dec. 9, 1822, for the New York Navy Yard; Nov. 6, 1822, for copper nails for the Boston Navy Yard; Jan. 15, 1823, for the *Enterprise;* Feb. 14, 1823, for copper to be sent to Portsmouth, N.H.; Feb. 18 and May 3, 4, 1823, for contracts for copper to be delivered to the Navy Yards at Gosport, Va., and Philadelphia; May 25, 1823, for bolts sent to navy agents at Norfolk; Aug. 5, 1823, copper for repairs to the *Guerrière.* On Aug. 8, 1824, copper was supplied for the schooner *Shark* and the *Constitution.* S. I. Isaacs & Soho Copper Co. with Commissioners of the Navy: Copper at Portsmouth, N.H., Washington, Gosport, Va. & Increase. Contract, Nov. 24, 1824; Naval Commissioners, National Archives and Records Service, Record Group No. 45, 338–339.

On Apr. 1, 7, 1825, a contract was negotiated for a sloop of war at Portsmouth and for the "General Increase" at Washington, D.C., May 4, 6, 1825, "copper to Charleston near Boston" and a sloop of war at the Brooklyn Navy Yard and one at Gosport, Va. On June 28 and July 12, 1825, various copper contracts were entered into with James K. Paulding, navy agent. These listings only comprise a small part of Hendricks' naval work. S. I. Isaacs & Soho Copper Co., with Commissioners of the Navy: Copper, Portsmouth, Boston, N. York, Philad., Washn. & Gosport. Contract, Mar. 31, 1826; Naval Commissioners, National Archives and Records Service, Record Group No. 45, 340–342.

In a descriptive letter to Mather Parkes & Co., Apr. 14, 1822, Hendricks wrote of the number of men in the copper import trade and the manner in which he frequently sold sheathing. Walter Harvey Weed, "Copper Deposits of New Jersey" in *Annual Report of the State Geologist* . . . (Trenton, 1903), 125, "In 1824 an expert smelterman was brought over from Germany and installed and worked a smelter near Bound Brook. A few years later he operated a small furnace near Belleville, N.J., the ore coming from the Schuyler Mine which was worked at intervals for the remainder of the century."

The *Waste Book* accounts of ships coppered by HH from 1823 on, state the source of copper used. In the letter to Mather Parkes, cited above, HH notes that Freeman copper is known "to have run 9 years."

Harris, *The Copper King*, 183, states that Newton, Lyon & Company acquired the Flintstone Mill between 1811 and 1813. They later traded as Newton, Keates & Company. Early in 1822 they offered their services to HH but this correspondence was not located. HH to Newton, Lyon & Co., June 10, 1822, and Newton, Lyon & Co. to HH, Liverpool, Aug. 20, 1822. According to Tooker, *Nathan Trotter*, 246, the company was in the copper and lead manufacturing business as early as 1805 and acted as brokers for the Marquis of Anglesea mines. Beside the stamp of "Mona" used by Anglesea, they stamped their ingot copper NL & Co. Mather Parkes & Co. were informed by HH, Sept. 5, 1822, that copper on board the *Meteor* and the *James Monroe* was not unloaded owing to the fear of yellow fever. Schreiber, Hoffman & Hulmes, London, May 13, 1826, reported the Lancashire riots, the depressed wool trade, and the inflated copper prices, conditions which contributed to increased immigration from England which in turn spurred the growth of the shipbuilding industry.

William H. Shaw, *History of Essex County* (N.J.) Vol. 2 (Philadelphia, 1884), 890 d–e, briefly recounts the enlargement of the mill in 1824 and the building of the Upper Mill in the settlement known as "Montgomery." Chauncey M. Depew, *One Hundred Years of American Commerce* . . . (New York, 1895), 333–334 adds this description of the mill in its early years: "Some of the buildings were of brick, roofed with tiles imported from Europe. The rolling mill was of wood, and contained one pair of breaking-down rolls, one pair of sheet rolls, and one pair of bolt rolls, all of which were imported from England." For other details on Soho see Case XVI, File 7. According to Depew, 334, the Gunpowder Copper Works of Baltimore was founded by Levi Hollingsworth in 1817, but Kauffman, *American Copper & Brass*, 27, establishes 1805 as the correct year of its founding. Thomas F. Gordon, *A Gazetteer of the State of New Jersey* (Trenton, 1834), 99, briefly notes the Belleville mill and other industrial operations in the area.

Hendricks' New England customers who commented on copper were: William T. Grinell, New Bedford, Sept. 25, 1828; Wyer Noble & Co. to HH, Portland, Me., Oct. 5, 1828, and Mathew Crosby, Nantucket, Apr. 29, 1829. Harrison & Sterrett to HH, Baltimore, Nov. 20, 1820, commented on McKim and conditions in their city.

IV

Family interests and activity are represented by the manuscripts in Case XII, Files 11 to 13, and in the correspondence of Frances Hendricks and Solomon I. Isaacs. Pool, *An Old Faith in the New World*, provides all of the detailed background for Congregation Shearith Israel, here and in subsequent references. The yellow fever and the growth of Greenwich Village as part of New York's earlier urban expansion have not been previously associated. Moses Levy Maduro Peixotto (1767–1828) was the forerunner of an illustrious family involved in medicine, journalism, foreign politics, Jewish affairs, and belles-lettres. The commonplace book of Selina Hendricks is in the Hendricks Collection. *Waste Book*, Oct. 2, 1821 and May 20, 1822 under miscellaneous accounts for "Rabbi Mosha." Case XI, File 10 Sept. 7, 1821, for Peixotto's purchases of old copper.

An objective examination of the social milieu of early American Sephardic Jewry is nonexistent. Nor is their any study of their rise and decline in the nineteenth century. Hendricks' attitude to Isaac Gomez, Jr. is boldly revealed in his letter to Gomez on July 9, 1816. Gomez was the author-editor of *Selections of a Father for the Use of His Children. In Prose and Verse*. (New York, 1820). In addition to numerous attestations of its literary merit, some editions carry a special word of praise from John Adams, "late President of the U.S." Hyman B. Grinstein, *The Rise of the Jewish Community of New York, 1654–1860* (Philadelphia, 1945), 144–149, where some of the charitable undertakings are discussed as well as the work of the *Kalfe*. Grinstein, 139–141, also reviews Hendricks' involvement in the *kasher* food problem. For Hendricks' contribution to Seixas' second

school see Wolf-Whiteman, 336. The contributions to Shearith Israel
are usually found under profit and loss in the ledger accounts. Grinstein, 40–42, and Israel Goldstein, *A Century of Judaism in New
York* . . . (New York, 1930), 56–57, for the rise and support of Bnai
Jeshurun. The Ledger account, Jan. 22, 1827, "to 5000 dollars loaned
this corporation 5 years @ 1 pc per ann." Case XII, File 11, contains
a typescript of the correspondence from the New Orleans congregation seeking a loan of $6000 from Hendricks, Aug. 27, 1829. *Occ. II*
(1844), 144 reports the gift of $100 in 1835 to the first organized con-
gregation in Cincinnati, O. *Proposals for Printing, By Subscription* . . .
*the Hebrew Pentateuch with the literal translation of each word in
English* (New York, 1826). Hendricks' name appears first among those
recommending the publication. Solomon Henry Jackson, *The Form
of Daily Prayers, According to the Custom of the Spanish and Portuguese Jews* . . . (New York, 1826).

Isaac Leeser, *Discourses, Argumentative and Devotional* . . . Vol. I
(Philadelphia, 1836), under "List of Subscribers," the Hendricks fam-
ily purchased five sets. See personal papers of HH, Case XII, File 16,
for subscription to Vanderlyn's *Rotunda* in 1818 and similar activities.
Tobias I. Tobias to Samuel I. Tobias, London, Aug. 2, 1815. Uriah
Hendricks, the son of HH and Mathias Gomez, first appeared in the
Waste Book in 1821 as mill employees.

V

The loose agreement of the original Solomon I. Isaacs and Soho
Copper Works, intended to terminate in four years, is perhaps the
only explanation for the ease with which the business was reorganized, the Isaacs name eliminated, and the two older sons admitted
as shareholders. Fortunately Hendricks took unconventional liberties
in the Ledger entries for the period 1827–1830, which yielded the
following: "Nov. 25, 1827—Advertised the Dissolution of the firm of
S. I. Isaacs & Soho Copper Company by an agreement made this
Day, the firm to be "Soho Copper Co." with the same parties as above
and the proffits to be divided as follows for Foreign and Home Business." For background to this move the entry of April 1, 1827, supplies
other details. Within the three-year period he noted that 2,109,488
pounds of raw copper were manufactured at an average cost of $19\frac{7}{8}$
cents per pound. Sources on the shipbuilders and copper companies
have been previously cited. Ship coppering records are found in detail in the *Waste Book* and in the *Ledger*.

Hendricks' work with and for the Hartford Bank contains some interesting banking history. Aside of the fact that the cashier Horace Burr always ascertained that messengers would not trespass on the Jewish Sabbath, or ask Hendricks to write, thereby violating the day, James Ward, a fellow director, could inoffensively jibe HH about a "Jew Banker" for a nunnery in his letter of Mar. 29, 1825. Hendricks was of importance to the Hartford Bank because of his intimate connections with the major banks of New York. In 1830, when the Hartford Bank hesitated to advance funds to the Hartford Fire Insurance Co. to cover major losses by fire, Hendricks advanced the money to the insurance company saving it from possible disaster. The details which are not in the Hendricks Collection are explained by Hawthorne Daniel, *The Hartford of Hartford* . . . (New York, 1960), 58. Gabriel P. Disosway, *The Earliest Churches of New York* (New York, 1865), 98. Walter Barrett, *The Old Merchants of New York* (New York, 1862), 117–118, wrote "that the credit of this house for half a century has never been questioned, either in this country or in Europe, and today in Wall Street, their obligations would sell quite as readily as government securities bearing the same rate of interest." *HHL II*, Jan. 6, 1818, $20,000 for bond and mortgages to John R. Livingston; Mar. 6, 1818, $20,000 to Isaac Moses and other members of the family. On Aug. 16, 1835, a £10 contribution was forwarded to Jerusalem through the offices of Nathan M. Rothschild.

Harmon Hendricks' retirement and reorganization of the business are recorded in the ledger as follows: "An account of the whole expense of the late Soho Copper Company which was formed on the first day of April 1827 between Harmon Hendricks and Henry Hendricks—and dissolved on 1st day of April 1830. The Capital invested was on Credit and on Interest at Seven pr Cent the whole amount advanced by Harmon Hendricks and reduced by him voluntarily to Six per Cent pr Ann." The final accounting of the mill for the three-year period follows.

5. Copper in an Age of Transition

I

The history of the uses of steam and the engines that made possible its application has been the subject of innumerable studies. Although its adaptation to the inland waterways held precedence over the locomotive, application of steam to land vehicles was a concomitant development. English railroad background is in L. T. C. Root, *The Railway Revolution* (New York, 1962). *Baldwin Locomotive Works. Illustrated*

Catalogue of Narrow Gauge Locomotives (Philadelphia, 1897), 7–50, covers the period relevant to Hendricks & Brothers' involvement in locomotive building. Interesting background on the manufacture of locomotives is in Joseph Harrison, Jr., *The Locomotive Engine and Philadelphia's Share in its Early Improvements* (Philadelphia, 1872).

H & B's surviving records are meager in comparison with those of their father. Their industrial activity has not been fully reconstructed owing to many obvious gaps in their papers. Documents relating to the reorganized firm from 1831 to 1869, the year of Uriah Hendricks' death, are in Case XVII, Files 1 to 10. The release signed by Solomon I. Isaacs, Feb. 1, 1831, and the *Articles of Copartnership* of Uriah, Henry, Montague, and Washington Hendricks, July 18, 1831, date the official beginning of the reorganized firm. H & B *Account Book* for the Belleville, N.J., mill is the source for the improvements to the mill, the administrative changes and procurement practices.

Typical of their frequent advertisements in the press is the following from the *NY. Enquirer*, Mar. 2, 1832.

Braziers Copper, Block Tin, &c.

The subscribers offer for sale on the most reasonable terms

20,000 lbs. Braziers Copper, assorted sizes and weight

10,000 do Raised and Flat bottoms from 12 to 34 inches

10,000 do Bolt Copper, assorted ½ to 1½ inch

20,000 do Banca Tin

5,000 do Composition Sheathing Nails

They also manufacture to order, Copper of every description and weight

Hendricks & Brothers   37 Beaver St.

"First quality American sheet copper" was offered to Joseph King, Jr., of Baltimore, Md., Oct. 21, 1831. Solomon Moses' correspondence in possession of M. W. stresses the continuity of his brokerage and commission business with the Hendricks family. The last known letter is dated May 5, 1843. The Joseph Lyons Moss correspondence with H & B, also in the possession of M. W., concentrates upon the locomotive builders stressing their dependency upon the copper works in Belleville. Its period of interest falls between Apr. 30, 1841, and Jan. 3, 1848. B. & I. Phillips were English Jews who settled in Philadelphia in the 1820's. They were brokers and commission merchants and advertised their goods and services regularly in the *Phila. Gaz.*, Jan. 2, Feb. 1, 10, Mar. 21, 1834, *passim*. They acted on behalf of Charles Rothschild of Paris, *U.S. Gaz.*, Feb. 16, 1836, and represented

Nathan M. Rothschild of London, June 27, 1834, *MS*, Lyons Collection, American Jewish Historical Society. For their sales of pig tin to H & B, see *Letter Book*, Jan. 19, 1831. For Leaming Brothers' pig tin accounts, Oct. 25, Nov. 16, 1831. Francis M. Drexel, founder of a major banking house, was first a portrait painter and then entered the commission business. He appears in the *Letter Book*, Aug. 31, 1831.

The regional spread of the use of copper and other nonferrous metals was evident in the 1820's; while H & B inherited their father's customers they made every endeavor to accommodate themselves to a new generation of metal craftsmen engaged in the new applications of copper. One of these, for example, was the manufacture of fixtures, chandeliers, and lamps for illuminating gas. The New York Gas Light Co. was supplied by H & B, *Ledger*, 1833–34. Solomon I. Isaacs reappears in the H & B accounts of his nephews, Sept. 7, 1831.

DAB, XIV, 525–526, for Anson Greene Phelps; Robert Glass Cleland, *A History of Phelps Dodge 1834–1950* (New York, 1952) for the emergence of the second largest producer of American primary copper. Although no documentary material on the Hendricks firm was found in the files of Phelps Dodge Corporation, Elisha Peck and Anson G. Phelps appear in the Ledger of HH prior to the organization of Phelps Dodge in 1834 and in HH accounts current for 1822, Case XI, File 12.

The transportation picture presented by McMaster, Vol. VI, 87–84, is upheld by the evidence in the Hendricks Collection. In 1833 plate and boiler copper for locomotives was supplied to the West Point Foundry Association, the Pontoosac Manufacturing Co. of Pittsfield, Mass., and Harrison & Sterrett of Baltimore. Fragments of reports of the Schenectady, the Western Railroad Corporation of Mass., and others are in Case XVII. Specifications for a double engine for the Mohawk and Hudson Railroad are dated June 5, 1840. *HHL II* under Stocks of Every Description shows the extent of the older Hendricks' investments. On Oct. 14, 1822 HH offered a loan of $65,000 to the Commission of the Canal Fund at Albany at $7\frac{81}{100}$ percent interest. Joseph & Co. entered into arrangements for Rothschild on the canal loan, Nathan M. Rothschild to the President and Directors of the New York Canal Fund, June 14, 1834, letter in the possession of M. W. James K. Medberry, *Men and Mysteries of Wall Street* (Boston, 1870), 13, discusses the canal loans of 1834–35 and comments, "There was a Jew operator, for example, who made occasional bank deposits of $500,000 in a day and men spoke of it under their breath."

Ledger accounts and correspondence files for 1833 indicate the extent of H & B business activity. Henry Hendricks' trip to Philadelphia, Baltimore, and the South is described in his letter, written from Baltimore, Oct. 22, 1836; Uriah Hendricks-Scovill correspondence from 1833–35. Scovill's plant is described in Whitworth & Wallis, *The Industry of the United States in Machinery, Manufactures and Arts . . .* (New York, 1854), 119–120. *Brass Roots. The Scovill Manufacturing Co.* (Waterbury, 1952), provides only a brief history of the Co. The Brainard and the Crocker correspondence is in the collection of M. W. Whitworth & Wallis also provide an early view of the New England mill trade.

II

Valentine, *Manual* (1865), 552, for Uriah Hendricks Levy's apothecary, *PAJHS* Vol. 17 (1909), 198 for the death of Gomez in a duel fight. *The Family Circle and Parlour Annual* (New York, 1850), 184 and 215, for examples of Uriah Hendricks Judah's literary contributions. For the literary work of Benjamin S. H. Judah, Lyle H. Wright, *American Fiction, 1774–1850* (San Marino, 1948), 160, and A. S. W. Rosenbach, *An American Jewish Bibliography* (New York, 1926), 190, 214, 244, and 288. The literature of the De Leon family is extensive and began to appear at this time. *DAB* Vol. V (1930), 224 for Thomas Cooper De Leon and Edwin De Leon.

The Hendrickses and their extensive family connections are important for the examination of the social, cultural, and business ties for this period. Barrett, *The Old Merchants*, offers additional family background in his frequent references to the old Jewish families of New York in the first three series of his informal accounts, although they are not fully reliable. All of the Jewish philanthropic societies that commanded the attention of these families involved Congregation Shearith Israel with which they were associated and they are discussed by Pool, *An Old Faith*, 358–366.

Silas Wright to Martin Van Buren, N.Y., Mar. 21, 1837 "rejoicing at the failure of the Jew brokers," *Van Buren MSS*, Library of Congress. Severe difficulties faced the Josephs, wrote Mrs. Tobias I. Tobias to Henry Tobias, N.Y., Jan. 15, 1846, "they are at their lowest ebb, Joe had his face slapped & hat knocked off in Wall St., he picked it up and pocketed the affront." McMaster, Vol. VI, 394–395, refers more fully to the Wright-Van Buren letter and also enumerates the business failures.

Frederick M. Tobias to Charles Tobias, Liverpool, Apr. 7, 16, 1838, on the depressed economic conditions in England and on the death of Michael Tobias. Henry Tobias to Charles Tobias, Liverpool, Apr. 27, 1838 on the watch trade.

Case XII, File 17, contains the medical records, the funeral details, and all of the legal items pertaining to the estate of Harmon Hendricks. His obituary appeared in the NY. *Commercial Advertiser*, Apr. 4, 1838. The figure of more than one million dollars is computed from the estate records. Barrett, *Old Merchants, First Series*, 116–117, writes that "He died immensely rich, leaving over three millions of dollars. His real estate was in the Sixth to Seventh avenues, from Twentieth to Twenty-second Streets; also out near where Mayor Wood lives, thirty acres on Broadway. Had he lived until now, as he never sold property, he would have been worth twelve millions." To what extent these estimates are accurate or speculative has not been determined.

Morris Tobias was one of the men who developed and manufactured chronometers. He was the uncle of the New York Tobiases. Frederick was the son of Michael J. from Liverpool and Henry the son of Tobias I. Tobias of New York. They conducted an extensive watch business with offices in New York, Philadelphia, and New Orleans, selling the popular Tobias patent lever watches. A large collection of Tobias papers is in the possession of M. W. After the death of Michael J. Tobias the company continued in the hands of his sons. See the *New Orleans Bee* and the *Commercial Advertiser*, both for May 23, 1838. T. I. Tobias, four of whose children married four of Harmon Hendricks' children, was initially in the linen and dry goods business and later entered the wine and brandy trade, establishing one of the noted houses in New York, Barrett, *First Series*, 117, for a brief description.

Correspondence between the English and the American Tobiases that made note of the now famous packets is in the letter of F. M. Tobias to Henry and Charles Tobias, Liverpool, May 23, 1838, "The Great Western has arrived here after a run of only fifteen days." On the same date to the same recipient, "Your letter pr. the Serius is safe at hand." And on May 28, 1838, "Your valued favors pr. Serius, Great Western, Virginian and North America all at hand." C. A. Acaryo (?) to T. I. Tobias, (?) June 28, 1839, comments on the ship's food for Jews observing the dietary laws. Charles to Henry Tobias, Liverpool, Feb. 16, 1839, on the captain providing *kasher* food. Henry to T. I. Tobias, Manchester, Dec. 8, 1836, forwarding a cheese, the "best that is made in this country" by a swift sailing packet, to be shared by the family. Rosalie (Roselane) Hendricks Tobias, born

Apr. 18, 1820, married Oct. 2, 1839, died Aug. 14, 1840. Harmon, Uriah's eighth child was born June 22, 1840, and died Mar. 9, 1841. Washington, born Nov. 22, 1807 died Mar. 16, 1841. He was unmarried.

III

Mathias W. Baldwin to H & B, Philadelphia, June 12, 1839, expressing his intention to meet with his creditors and asking H & B to wait for better times. H & B obviously consented to this request. This is one of twenty-four letters of Baldwin under various business names in the possession of M. W. Baldwin & Vail to H & B, Philadelphia, Dec. 21, 1841 on the default of the Michigan Rail Road.

Deposition of Charles A. Gurley, Sept. 10, 1840, on the Detroit and Pontiac Railroad. Gurley was on the staff of H & B. An account of William Norris is in Harrison, *The Locomotive Engine*, 37–39. The Norris letter to H & B, Jan. 13, 1841, where he writes of the order for twenty-six locomotives and his guarantee to build a first-class engine, Philadelphia, Aug. 8, 1843, is in the possession of M. W.

The letters of Joseph Lyons Moss to H & B that reveal many of the confidential aspects of locomotive building and sales in Philadelphia are dated as follows: July 12, Aug. 11, 18, 1843. These and the Harley papers are in the possession of M. W. Metals reaching the port of Philadelphia and the activity of the Taunton Copper Works agent, are reported by J. L. Moss, Apr. 30, Nov. 19, 1841, and Jan. 6, 1842. Charles Tobias to Uriah and Henry Hendricks, New Orleans, Jan. 2, 1843. Henry Hendricks to Henry Tobias, New York, May 1, 1844, on the annexation of Texas.

IV

J. W. Foster and J. D. Whitney, *Report on the Geology of a Portion of the Lake Superior District* (Washington, 1850), published as a government document, is the first major investigation of the first major copper producing region in the United States. Thomas H. Blake, Commissioner, General Land Office, to H & B, Washington, D.C., July 18, 1842 in reference to their claim on Michigan Public Lands. William B. Gates, Jr., *Michigan Copper and Boston Dollars* (Cambridge, 1951), 1–38, traces the economic development of the mining region.

Whitworth & Wallis, 118, describes the G. H. Hussey copper smelting works and the manner of conveyance from mine to mill. Receipt from William Bingham to H & B, N.Y., Apr. 23, 1844, for 6,162 pounds of copper shipped to Cincinnati at 75¢ per 100 pounds. More than 345 buildings were destroyed by fire in 1845 at an estimated value of $6 million. George M. Tobias to T. I. Tobias, Liverpool, Aug. (?), 1845 speaks of the miraculous escape of T. I. Tobias from the fire.

Henry Hendricks to Henry Tobias, N.Y., May 1, 1844, describes the *bar mitzvah* preparations for Isaac, his oldest son. Of the four congratulatory notes received by Joshua, a son of Uriah, T. I. Tobias' letter, Oct. 22, 1845, is affectionate and rich in spiritual counsel. HH's vast real estate holdings were advertised extensively in the press and by huge auctioneer's posters which are in the Hendricks Collection. Mrs. T. I. Tobias in her gossipy letters to Henry Tobias reported on Jan. 15, 1846, the behind-the-scenes activity and could not refrain from adding that, "nothing has happened but hard times, War, War, Peace, Peace, stocks down and up." After the sale Harriet Hendricks Henry wrote to Henry Tobias, N.Y., Feb. 9, 1846, describing some of the real estate acquisitions and the plans of Benjamin Nathan to build at Belleville.

The remodeling of the New York offices and warehouse is described in the *Articles of Agreement* between Otis Pollard and H & B, Jan. 20, 1846. The correspondence of Merrick & Towne, known also as Merrick & Sons, with H & B is in the possession of M. W. It relates to the building of the *Mississippi*, the *San Jacinto*, and difficulties in producing the proper copper. They are dated, Philadelphia, Jan. 17, Feb. 14, 15 and Nov. 7, 1848. Scharf and Westcott, *History of Philadelphia*, Vol. III (Philadelphia, 1884), 2253, lists these ships among others. Charles B. Boynton, *The History of the Navy During the Rebellion* Vol. I (New York, 1867), 242, for the principal screw war steamers of the Navy. Contract, H & B and the United States Navy Department, June 26, 1851, to furnish and deliver to the Brooklyn Navy Yard 200 pounds of India tin, 800 twenty-pound sheets of braziers copper, 16,000 pounds. The Belleville White Lead Co. was founded Apr. 19, 1847, and in 1860 it was leased to H & B. New Jersey Oil Co. vs. H & B, Dec. 1, 1853. Judging from the surviving stock and bond certificates, H & B's investments support their tremendous interest in mining, minerals, and transportation. Shares were purchased in the New York & Havre Steam Navigation Co. on June 15, 1850; in the Commercial and Rail Road Bank of Vicksburg, Miss. on June 25, 1844; 313 shares at $100 each were purchased of the Sewanee Mining Company, Nov. 22, 1854, and The Union Consolidated Mining Co. purchased in the 1860's. Both of the latter were in Tennessee. The Hamilton Copper Co., the Isle Royale Mining Co., the Knowlton

Mining Co., all of which were in Michigan are represented by shares in the H & B papers. Medberry, *Wall Street*, 274–277, writes of the extensive investments in mining securities and the organization of a mining board.

V

Although minor errors are present in the genealogical chart of Uriah Hendricks' family, compiled by Stern, *Americans of Jewish Descent*, 80, it presents a clear picture of this extensive branch of the family and its collateral members. Uriah Hendricks to Joshua Hendricks, New York, Aug. 31, 1852. Adelaide married Thomas Jefferson Tobias in 1853. This Tobias family settled in Charleston prior to 1737: Barnett A. Elzas, *The Jews of South Carolina* (Philadelphia, 1905), 24. On Oct. 29, 1855, T. J. Tobias suggested a partnership with Joshua in the general commission business. Joshua's *Letter Book* 1856–57 shows that he entered the brandy import trade but details of his business and other activity cannot be defined owing to the badly faded letters. But some of the deciphered correspondence, Jan. 17, June 2, 1857, shows that he was importing copper, tin, and spelter from Newton, Keates & Co. of Liverpool, a firm with whom his grandfather had entered into business shortly after the War of 1812. He married Emma Brandon, Jan. 28, 1857. The Brandon papers are also in the Hendricks Collection at the New-York Historical Society and contain Joseph Brandon's financial records.

Uriah Hendricks Judah was the author of the obituary of Naphtali Judah which appeared in the *Occ.*, Vol. XIII (1855), 420. Accurate population figures for the Jews of New York City are not known for this time. Grinstein, 469–471, projects the best data thus far available. He arrives at the figure of 40,000. Whatever its accuracy, the number of Jews was great enough to stimulate considerable institutional growth. *Constitution of the Jews Hospital in New York* (New York, 5612 [1852]) for members of the Hendricks family. Joseph Hirsh and Beka Doherty, *The First Hundred Years of the Mount Sinai Hospital of New York* . . . (New York, 1952), for a fuller account of the Hendricks' participation in its affairs. The *Asmonean*, Aug. 21, 1857, announced the work of Isaac Hendricks and Jacques Judah Lyons, the minister of Shearith Israel, in raising funds to be transmitted for the Geneva, Switzerland, synagogue. Isaac B. Kursheedt and others to Sir Moses Montefiore, no place, no date, retained copy in Leeser Collection, Dropsie University, Philadelphia, transmitting $300 for the "Poor of our Brethren in the Holy Land" through Phelps, Dodge &

Co. Selina Hendricks advertised in the *Occ.*, Vol. V (1847) 3 (advertising section), that she taught pianoforte, guitar, etc. Gustavus Isaacs to Isaac Leeser, New York, Dec. 16, 1855, transmitting $100 in the names of Selina Hendricks and Hannah Isaacs for the graveyard and synagogue in Lancaster, Pa. Leeser published a critical appraisal of the attitudes of Sephardic Jewry in the *Occ.*, Vol. XVII (1859), 223, and his cooperative efforts that enabled the Hendricks family to dispense charity is further indicated in the letter to him by L. Baum, Lancaster, Pa., Aug. 12, 1859, requesting additional aid from the Hendricks family.

VI

Receipt for subscription by H & B to the New York Exchange Co., May 21, 1858. Their precise activity with the Exchange has not been determined. Francis Hendricks (1836–1912) received a Master of Arts degree from Columbia College in 1856. It was he who eventually entered into the business of commission merchant with the Charleston Tobias. *Articles of Agreement* made the first day of June, 1859, between T. Jefferson Tobias (who had come to New York) and Francis Hendricks of New York and Joseph L. Tobias of Charleston, S.C. The agreement bound them for three years.

Articles of Copartnership made and entered into June 17, 1859, between Henry S. Henry, Solomon de Cordova, and Isaac Hendricks all of the City of New York for a term of three years. The extensive correspondence of this firm, which includes Texas material, is in the Hendricks Collection. Only fragments survive on the commercial activity of Mortimer Hendricks, who was a partner in the firm of Garbutt, Black and Hendricks. Shaw, *History of Essex County*, 890 d–e, notes the improvements to the mill and the manufacture of rivets and copper wire.

VII

Mordecai Hyams of Charleston, S.C., who worked for Tobias, Hendricks & Co. arranged on their behalf and fifty New York firms to engage a schooner to sail for Cuba loaded with whatever real assets that could be salvaged from the fate of war. By 1863 Tobias, Hendricks shifted to wholesale liquor merchandising. Bond coupons of the Florida Rail Road Co. in stock and bond certificate file. Henry Hendricks died at the age of fifty-one, *Occ.* Vol. XIX (1861), 336.

Marcosson, 50–56, reviews the military uses of copper and the contributions of the New England industries in supplying metals to the

Union Army, but like most historians was unaware of the work of H & B. Edmund Hendricks was a private in Company F, 7th Regiment, New York State National Guard, and though he did not see active duty he continued as a member of the Regiment until Apr. 24, 1872. Uriah Hendricks' name appears on the printed circular of the Sanitary Commission, Central Finance Committee, N.Y., June 26, 1861. Pool, *An Old Faith*, 60, 333, 367, refers to the Ladies Army Relief of Shearith Israel.

The miscellaneous correspondence in H & B file and the *Letter Book* of Joshua Hendricks, Apr. 6, 1863, contains letters to Edmund. Letters of Joshua to Florian Florance (who married Sarah Hendricks), on Isle Royale copper, Feb. 4, Apr. 7, 1863; the stock of the New Bedford Copper Co., Feb. 19, 1864, and Joshua to Edmund Hendricks on the demand for Silesian spelter. In his own correspondence Uriah wrote to Edmund, N.Y., May 15, 1862, advising him on the price of spelter, antimony, and the purchase of Silesian metals. MacLean Maris to H & B, London, Nov. 3, 1863, acknowledging an order for twenty-five tons of spelter and informing H & B that tin and tin-plate are scarce, prices have risen and lead is more expensive. Gates, *Michigan Copper*, 15, on demands made by the war. George Edgar Turner, *Victory Rode the Rails* (Indianapolis, 1953) for an overall view of Civil War railroading. J. Edgar Thomson to Enoch Lewis, no place, Mar. 23, 1864, "introducing to you my friend Mr. Hendricks of New York," H. W. Schotter, *The Growth and Development of the Pennsylvania Railroad Company* (Philadelphia, 1927), 33 for John Edgar Thomson. Atlantic & Great Western Railway Company requesting the presence of H & B at the opening of their connection between the Atlantic seaboard and Cleveland, O. The 1864 balance sheet, June to May, records entries for Baldwin Locomotive Works, Schenectady Locomotive Works, New Haven Arms Company and the Watervliet Arsenal. Joshua Hendricks *Letter Book*, to Neafie, Levy & Co., Apr. 3, 1863, inquiring if they can build "2 iron lights."

Shares in the Benton Gold Mining Co. of Colorado, organized April, 1864 and the Central Gold Mining Co. of Colorado, organized about the same time, appear in the stock and bond certificate case of H & B. Railroad stocks are registered in the books of Joshua Brandon.

Jewish Encyclopedia Vol. IX (New York, 1916), 177, contains an accurate account of Nathan's financial, social, and Jewish interests but does not state that he was murdered under mysterious circumstances.

Joshua was an active purchaser of Ridge Copper Company Mining stock from 1863 to 1873; Edmund, of Minnesota Mining Company in 1866; Francis of Isle Royale of Michigan in 1863, and Uriah Hendricks of The Mariposa of California in 1866 and New York and Nova Scotia Gold Company in 1863.

Joshua's trading methods are revealed in the memoranda of Lee and Danforth of Boston, Feb. 19, 1873, and others. He held shares in the Lake Superior Mineral Land Company, which were acquired through William Z. Florance. These investments represent only a small segment of the shares held in eighteen states.

Ingot copper used as collateral is noted in the letter book of Uriah Hendricks, April 10, 1865. A certificate from the Mechanics Bank dated June 10, 1867, records 400,000 pounds and on Aug. 5, 1869, after the death of Uriah Hendricks, the collateral was increased to 500,000 pounds of refined copper. This instrument empowered the Mechanics Bank to sell at public or private sale whatever amount of copper necessary, if the account of H & B was overdrawn.

Balance sheet dated May 1, 1864, U.S. Ordnance, $8,400; July 1, 1864, $25,391.25, and metals for the U.S. Bureau of Engraving, $2,262.50. The sales for the above period totaled $691,976.03.

Lazelle Perkins & Co. to J. C. Hoadley, N.Y, July 14, 1865, written by E. W. Barston to the New Bedford Copper Co., the letter contains the allegations that H & B, Revere, and Crocker were price cutting and thereby undermining the purpose of the copper association. Charles S. Randall to H & B, New Bedford, July 18, 1865, forwarding a copy of the above letter. H & B to E. W. Barston, N.Y., July 17(?), 1865 replying to the charges.

Bertram W. Korn, *American Jewry in the Civil War* (Philadelphia, 1951), 112–113, reviews the steps taken in the North to supply Jews in the South with Passover food. Solicitation of funds began after the fall of Savannah to the Union Army, thus on Mar. 3, 1865, Uriah Hendricks wrote to Isaac Leeser that he already made his contribution. Uriah to Mrs. Augusta Tobias (of Liverpool), N.Y., Aug. 4, 1865, on the return of peace.

Edmund, judging from his father's letter of Oct. 13, 1865, was in Richmond. On July 1, H & B supplied the Navy Office of Supplies 2,000 sheets of yellow sheathing for the ship *Onward*. Similar information on the metal trade is in the same letter. There are approximately 500 letters of Tobias, Hendricks and Company between May 27, 1865, when business was resumed and Apr. 3, 1866. They have not been studied.

Joshua to Edmund Hendricks, N.Y., Oct. 14, 1865, acknowledging orders for turpentine stills and forwarding $8,000 in gold for the purchase of southern goods. Edmund to Tobias, Hendricks, Columbia,

S.C., Oct. 16, 1865, on cotton prices, railroads, and contacts with the planters. Letters from Edmund on Oct. 19, 24, 28, 1865, describe his experiences with Joseph Lopez, Joseph Tobias, and Raphael J. Moses. Jacob R. Marcus, *Memoirs of American Jews* Vol. I (Philadelphia, 1955), 146–147, contains a fine summary of Moses. For Edwin De Leon see earlier note and Elzas, *The Jews of South Carolina*, 205, 227, and 273.

Constance Hendricks to Edmund, N.Y., Oct. 29, 1865, and Francis Hendricks to Edmund, N.Y., Nov. 6, 1865, chatting about the social life of New York Jewry. Solomon Cohen was introduced by Octavus Cohen to Tobias, Hendricks & Co., Savannah, Aug. 19, 1865. The Minis family which is related to the Hendricks family was and still is one of the prominent Jewish families of colonial Georgia background. Barry E. Supple, "A Business Elite: German-Jewish Financiers in Nineteenth Century America," in *The Business History Review*, Vol. XXXI (1957), 144–145 and 147 for James Speyer. Edgar McMullen to Edmund Hendricks, Charleston, Feb. 14, 1866, inviting Edmund to return South and become a planter.

VIII

Gates, *Michigan Copper*, 39 *passim* gives an important picture of copper mining, production, and distribution immediately after the war. Medberry, *Men and Mysteries*, 236, under the heading "Statement of Jew, German, & Co. to New York Gold Exchange Bank." The gold exchange issued undated circulars that listed the names of those who traded in the gold market.

Jewish congregations throughout the nation met to protest the action of the fire insurance companies. Particularly bitter were the Jews of Richmond, Va., who were joined by Joseph Mayo, the mayor of the city, Richmond *Times*, Apr. 8, 1867. In Philadelphia, Congregation Rodeph Shalom undertook to organize the national protest, Alex Reinstein to Abraham Hart, Mar. 18, 1867, Archives of Congregation Mikveh Israel, Philadelphia. Uriah Hendricks' faded letter book contains one letter, Mar. 8, 1867, that is vital to an understanding of the entire unpleasant episode.

Uriah Hendricks to F. W. Capen, Sec'y. Isle Royale Mining Co., Aug. 9, 1867, politely rejecting a directorship of the company. Gates, *Michigan Copper*, 38–63, discusses the period from 1867 to 1884. The doggerel which is unsigned is dated Apr. 29, 1868. William H. Michael and Pitman Pulsifer, *Tariff Acts Passed by the Congress of the United States from 1789 to 1895* (Washington, 1896), 232–233, for the act of

1869 "regulating the duties on imported copper and copper ores." Obituary notices of Uriah Hendrickš appeared in most of the New York press for the week of Mar. 25, 1869.

Epilogue

I

Joshua Hendricks, who became head of the new firm, revised the name of Hendricks & Brothers by eliminating the &. In 1874 it became the Belleville Copper Rolling Mills. Marcosson, *Copper Heritage*, 48, for the admission of outsiders to the Revere Company in 1828. Watson Davis, *The Story of Copper* (New York, 1924), 207, published a table of copper alloys used in the fabrication of various metals. Marcosson, *Anaconda* (New York, 1957), 12–29, on the discovery of precious and nonferrous metals. C. B. Glasscock, *The War of the Copper Kings* (Indianapolis & N.Y., 1935) is also useful.

By 1872 the offices and warehouse of Hendricks Brothers were located at 49 Cliff Street; see permit to receive and discharge goods from this location, May 1, 1872. These offices were maintained until May, 1939, when the business was finally liquidated. James A. Alexander for the Aetna Fire Insurance Co. of Hartford, Conn., Nov. 2, 1874, issuing a statement of loss by fire as of Oct. 12, 1874, on buildings, machinery, and equipment for $27,250. Shaw, *History of Essex County*, 890 d–e states that the "Upper Mill was destroyed by fire, and was immediately rebuilt in a larger, more beautiful and imposing scale, and presents itself as a magnificent manufactory." In 1875 the lower, or "Soho Mill," was destroyed by fire. It is interesting to note that despite the clash with the insurance companies, Hendricks Brothers continued as policy holders with the Aetna. Shaw also describes the "large tracts of land, with the spacious mansions and beautiful surroundings." The James Moore noted by Shaw is not to be confused with John A. Moore who according to Barrett, *Second Series*, 141–142, "had thoroughly learned the business with old Mr. Hendricks . . ." It may have been at the time of the rebuilding of the mill that "The largest rolls for copper in the country and probably in the world, 156 inches long and 30 inches in diameter" were installed. William G. Lathrop, *The Brass Industry in the United States. Revised Edition* (Mount Carmel, Conn., 1926), 45 makes this statement but does not indicate when the rollers were installed.

Gates, *Michigan Copper*, 45–46, on the tariff of 1869 and its aftermath. Marcosson, *Copper Heritage*, 68, gives some background on

Pope, Cole, but the problems confronting them which led to their absorption by the Baltimore Copper Rolling and Smelting Company in 1887 are not clarified. Pope, Cole & Co. to J. R. Magruder at Fort Bayard, N.M., Baltimore, June 5, 1874, accusing HB of breaking down the market price of copper. See *Copper Manufacturers' Association Constitution*, 1891, for Pope and Joshua Hendricks. Gates, *Michigan Copper*, 48, on Holmes and Lissberger. The option with Magruder is dated Feb. 15, 1875, for the Mimbres Mining Company. Some of the mines were the San Jose, Yosemite, and the Chino. The Company was incorporated April 3, 1880, and on April 5, John R. Magruder, its president transferred 3,000 shares at $10 each to HB. Located in the area of Grants, New Mexico, it became the Jackpile uranium mine of Anaconda.

S. Policoff to Grant Locomotive Works, Charkoff [Kharkov] Nov. (?), 1874, agreement for the delivery of 20 passenger locomotives of 8 wheels and 30 of 8 wheels on axles, and others to Taganrog, Russia. The Grant Works was the New Jersey Locomotive and Machine Company of Paterson, N.J., O. D. F. Grant, president; see *Once a Month* Vol. 2 (1867), 40. When Joshua Hendricks visited Paris he wrote on Nov. 7, 1892, about locomotive contracts in Philadelphia, that Baron de Hirsch held about 24,000 tons of fine copper and that the Rothschilds showed no interest in the copper market.

Edgar Hendricks (1852–1895) received his certificate from the New York Metal Exchange on June 18, 1884. Henry Harmon Hendricks (1859–1904) graduated from the Columbia School of Applied Sciences, 1880, *Catalogue . . . Graduates of Columbia University* (New York, 1916), 689. About Clifford Hendricks (1862–1901) little is known other than the activity indicated.

Shortly after the death of Montague Hendricks an agreement drawn on May 29, 1886, was entered into between the brothers, Joshua, Edmund, and Harmon Washington Hendricks that the interests of either one, except personal finances, would be retained by the firm in case of death.

II

Studies of social and professional clubs have taken for granted that Jews were not within the exclusive circles of midnineteenth-century society. It is important to recognize the participation of the Hendricks family and their kin in the various clubs, some of which a century later are being assailed for not admitting Jews. Although they did not share the compulsion or need to join prominent exclusive clubs, their names appear on the membership rolls of many of the New York and New Jersey clubs. Benjamin Nathan who married Emily Grace Hendricks was a member of the St. Nicholas Society, the Union and Union League clubs. *The New York Club, Officers and Members* . . . (New York, 1889), 23–24, lists five Hendrickses. It was the second oldest club in the city, founded in 1845. *Constitution and By Laws . . . of the New York Yacht Club Organized 1844* (New York, 1888), 43 lists three Hendrickses as members. It is interesting to add that the Philadelphia *Jewish Exponent*, Oct. 12, 1962, stated that Emil Mosbacher, skipper of the prize winning *Weatherly* that year, was the first Jew to be admitted to the Yacht Club "which never had a Jewish member in its history." This observation was typical of the Jewish and general press. *Constitution and By-Laws of the Fulton Club* (New York, 1889), 36, names Joshua as a cofounder and lists four Hendrickses as members. *Essex County Toboggan Club* (Orange, N.J., 1889), 13, lists five Hendrickses as members. This is only a partial list of the clubs in which they were members and which list many other Jews. Social and family gossip, some pertaining to club life, is in Frances Nathan Wolfe, *Four Generations* (New York, 1939), 21–52.

The vast literature relating to the East European Jewish immigration cannot be reviewed here. But the *Annual Report of The United Hebrew Charities of the City of New York* from 1874 to 1915 embody information relevant to the Hendricks family. *The Proceedings of the Seventy-Fourth Annual Meeting . . . of the Hebrew Benevolent and Orphan Society* (New York, 1898), which was organized to provide relief for aged Jewish soldiers of the American War for Independence, and the *Proceedings of the Conference of Hebrew Emigrant Aid Societies . . .* (New York, 1882) indicate the involvement of the Hendrickses in Jewish philanthropic work.

Raphael J. Moses, Jr., to HB, N.Y., Feb. 18, 1876, appealing for funds for the YMHA. The *N.Y. Times*, May 16, 1912, published the resolution of The New York Society for the Prevention of Cruelty to Children memorializing the years of devoted activity of Harmon Hendricks (1838–1912), one of the original incorporators of the Society in 1875 and its vice president for twenty-six years. Pool, *An Old Faith*, 392, for the Mount Sinai Training School for Nurses. The visit to Turin, Italy, was described by Blanche Hendricks, wife of the above Harmon, to Joshua, Turin, Mar. 28, 1877.

S. T. Snow, *Fifty Years with the Revere Copper Co.* (Boston, 1890).

It was Snow who forwarded the Revere-Hendricks Correspondence to Edmund the year before. Snow apparently enjoyed a friendship beyond commercial interests. On Mar. 26, 1875, he wrote Joshua asking for historical material on Jewish customs and rites for a book that he was writing.

The sale of timber lands is mentioned by Edgar to Joshua, N.Y., Oct. 25, 1892. Harvey O'Connor, *The Guggenheims* (New York, 1937), 63. *Who's Who in American Jewry, 1926* (New York, 1927), 383–384, and *JE* Vol. VIII, 69–70, for the Lewisohns. The Lewisohn Brothers, Leonard and Adolphe, were both born in Hamburg, Germany, and did not enter the New York metal trade until the 1870's. Correspondence between Lewisohn Brothers and HB, N.Y., Jan. 6, 19, 1899. Phelps, Dodge maintained offices in the New York area continuously since 1834; see Cleland, *A History of Phelps, Dodge*. William E. Dodge to Joshua Hendricks, N.Y., Apr. 13, 1891, on the New York Smelting and Refining Co.

For Joshua Hendricks' trip to France and England see correspondence for 1892. *The Shipping and Commercial List and New-York Price Current*, Dec. 22, 1894, on the celebration of its centennial, published a special article on the history of the five generations of the Hendricks family noting that they were still doing business with the colonial firms identified with the first Uriah Hendricks. Such accounts of the firm were seldom published. Even the *JE* which contains a genealogical record of the family does not include a history of the family.

III

Details of the death of Edgar who died in 1895 and Clifford who died in 1901 are not known. The *N.Y. Times*, May 28, 1904, reported the death of Henry Harmon Hendricks. Edmund Hendricks died in 1909 at the age of seventy-five. Francis was seventy-six at the time of his death in 1912. An intensive search for business records for this period proved fruitless.

Harmon Washington Hendricks' twentieth century work is associated more with the Museum of the American Indian than with any other interest. His contributions to the Museum were considerable. The most recent scholarly publication that recalls his name is Watson Smith, Richard B. and N. F. S. Woodbury, *The Excavation of Hawikuh by Frederick Webb Hodge. Report of the Hendricks-Hodge Expedition 1917–1923* (New York, 1966). An appreciation of Harmon W. is in the foreword.

IV

Henry Solomon Hendricks (1892–1959) was the son of Lilian Henry and Edgar Hendricks. All of the material used here is based on the personal papers in the possession of Rosalie Gomez Nathan Hendricks who made them available for this study. In addition to these valuable papers, Pool, *An Old Faith*, presents a picture of his role in Shearith Israel. *PAJHS* Vol. XLIX (1959), 56–63, for Pool's necrology and Edgar J. Nathan, Jr., "Henry Solomon Hendricks," in *The Association of the Bar of New York. Memorial Book* (New York, 1960), 36–38, for his legal activity. *PAJHS*, 1892–1932, contains numerous references to the participation and gifts of various members of the family.

The *Newark Evening News*, July 26, 1924, published an account of the buildings that stood on the tract and the gift of the park to Essex County. *N.Y. Times*, Dec. 4, 1928, contains the proceedings of the metal exchange. Smith, Chambers & Clare to Helen R. Hendricks, N.Y., Mar. 2, 1939, on the liquidation of the mill by public auction. The surviving records including Harmon Hendricks *Letter Book*, 1809–1825, were temporarily deposited with the Farmers Trust Company. In the process of transfer many of the papers were dispersed and are still circulating on the manuscript market.

In an unusual historical coincidence, the descendants of John Freeman & Co., Samuel T. Freeman & Co. of Philadelphia, cried the auction of what was once the oldest copper mill in the United States.

*A Selection of Documents
Relating to the History
of Early American Copper*

THE SOHO MILL

Extract from Report of T. P. Cope, Dated July 4, 1800.

"Took passage in the stage for Soho Works, near Newark, New Jersey, on the morning of the 3d of July, 1800, and arrived there about noon of the next day.

"Soho is named after the works of Bolton & Watt, in England, and is situated about three-quarters of a mile northwest of the Passaic, on a small stream called Second River.

"The works consist of, a smith-shop 90 x 40 feet, with six fires and two air furnaces; next to this is a room 30 x 20, in which is the fire, for heavy work; four wooden bellows play into a regulator 15 x 15 feet, with pipes to the forge, and four furnaces for melting and refining copper. Then there is a stone building 20 x 24, two stories high, with six stampers for preparing loam for the furnaces; next to this is a fitting shop with large lathe and drilling machine, and a water-wheel 20 feet diameter, to bore cannon; next to this is a shop with a water-wheel 30 feet diameter for boring large cylinders; this is now boring a small cylinder for a steamboat which belongs to Roosevelt, Chancellor Livingston, and others.

"Higher up the stream is the furnaces, 60 x 50 feet, with two air furnaces capable of melting 40 cwt. of metal each, two blast furnaces for melting and refining copper, with a coal house and pattern shop, with two foot lathes; all are stone buildings; the stream affords a head and fall of 16 to 18 feet."

[Fred Graff], *Notes upon the Water Works of Philadelphia, 1801–1815* (Philadelphia, 1876), 1–2.

The Soho Company, Incorporated the 27th November, 1801, By the Assembly of the State of New Jersey. New-York: Printed by T. & J. Swords, No. 99 Pearl-street. 1802.   .

State of New-Jersey

An ACT *to incorporate the Persons therein named, and their Associates, under the Name of* THE SOHO COMPANY.

Whereas Nicholas I. Roosevelt, by his petition to the legislature, in behalf of himself and his associates, hath set forth, that he hath obtained a lease for the copper-mine, situate in the county of Bergen, within this State, commonly known by the name of Schuyler's Mine; that, under the said lease, the company hath expended considerable sums of money, in erecting engines, clearing old levels and opening new ones, for the profitable working of the same; and that, although the said petitioner hath no doubt of the certain profits the mines would yield, yet he hath not funds any way adequate to the carrying them on; and praying that the legislature would pass an act, incorporating the said petitioner and his associates for the purpose aforesaid: Therefore,

1. *Be it enacted by the Council and General Assembly of this State, and it is hereby enacted by the authority of the same,* That Aarent I. Schuyler, Nicholas I. Roosevelt, John Stevens, Robert R. Livingston, Samuel Corp, and James Casey, and all others who are or shall become subscribers and associates for the purposes of this act, shall be, and they are hereby, together with their assigns and successors, established and made a body corporate and politic, for the purpose of conducting the operations of a copper-mine and metal company, and for no other purpose whatever, by the name and style of *The Soho Company;* and are hereby ordained, constituted, and declared to be, for and during the term of fifty years, a body politic and corporate, in fact and in name; and by that name they and their successors shall be, and are hereby made able and liable, in law, to purchase, take, hold, occupy, possess, and enjoy, to them and their successors, any lands, tenements, hereditaments, goods, chattels, and effects, of whatever kind; and the same to sell, demise, grant, or dispose of: also to sue and be sued, plead and be impleaded, defend and be defended, in any court of record and elsewhere. And the said corporation shall and may have and use a common seal, alterable at their pleasure; and by such seal, for the time being, their proceedings, deeds, and transactions, shall

and may be certified and established. Provided, that the whole estate of the said corporation shall not exceed, at any time, the sum of *half a million of dollars,* exclusive of the ores and metals which they may possess and occupy under any actual productive operation.

2. *And be it enacted,* That the said corporation shall and may, at a general meeting of the stockholders, to be called as herein-after is provided, and afterwards at such other times and places as shall be by them agreed and determined, appoint six directors and a treasurer, with such clerks, agents, and other officers, as shall appear necessary for conducting the affairs of the said corporation; and at such general meeting, or otherwise, as they shall agree, shall and may establish and put in execution reasonable by-laws, ordinances and regulations, with or without penalties, for the calling of future meetings, and for the government and management of the said corporation, and for the ordering, employing and restraining of their overseers, tenants, servants, artificers and labourers: provided that such by-laws and regulations shall not be repugnant to the constitution and laws of this State, and the United States.

3. *And be it enacted,* That the capital stock of the said company shall be divided and holden in shares, so that fifty dollars of such stock shall constitute one share; and in the meetings and transactions of such corporations, every stockholder, holding one or more shares, shall be entitled to one vote for each and every share he may so hold.

4. *And be it enacted,* That the first meeting of the said corporation shall be holden at Newark, at such time as a majority of the persons herein before named shall judge proper, after previous notice being given for at least two weeks in two of the newspapers printed in this State, and in one of the newspapers in each of the cities of New-York and Philadelphia.

5. *And be it enacted,* That the said corporation may be dissolved at the general meeting specially summoned for that purpose, provided at least three-fourths in value of the stockholders shall be present, or represented therein; and upon such dissolution, or upon expiration of the said term, the directors for the time being, and the survivors or survivor of them, shall be *ipso facto* trustees for settling all the affairs of said corporation, disposing of its effects, recovering and paying its debts, and dividing the surplus among the stockholders, in proportion to their respective interests in the stock, unless the stockholders, at such general meeting, or at the expiration of the said term, shall appoint other persons, not less than three, nor more than five in number, for such purpose; in which case the persons so appointed, and the

survivors and survivor of them shall be trustees and trustee for the purpose aforesaid.

Passed at Trenton, November 27, 1801.

The foregoing is a true copy from the original, examined and compared by

MASKELL EWING,

Clerk of the General Assembly.

The Schuyler Copper-Mine.

This mine, situated between the rivers Passaick and Hackinsack, near their confluence, in the State of New-Jersey, was discovered about the year 1719, by Aarent Schuyler, grand-father of the gentleman of that name now living. The ore was found on the side of the hill, and was easily raised; but as the policy of England at that time prohibited the establishment of smelting works, or manufactories, in her colonies, it was packed in casks, each containing about four hundred weight, and exported in its state of ore to England. It appears by his books, that before the year 1731 Aarent Schuyler had shipped 6,938 casks, making about 1,386 tons of raw ore, to the Bristol copper and brass company. His son, Colonel John Schuyler, prosecuted the work with more numerous and more skilful hands. The quantity of ore raised by him is not known, as his books were lost during the war. In 1761 the mine was leased to a company, who erected a steam-engine, of the imperfect construction then in use. The engine-house, constructed of combustible materials, was soon afterwards burned down. It was, however, rebuilt, and the mine was worked for four years with great advantage and profit. In 1765 a workman who had been dismissed, set fire to the engine-house. It was again destroyed, and the works were discontinued by the company. Several gentlemen in England, however, whose connection with the company had taught them the superior quality of the ore of Schuyler's mine, applied successfully to the crown for permission to establish works in America for smelting and refining copper; and an offer was made to Mr. Schuyler to purchase the whole estate containing the mine, for the sum of one hundred thousand pounds sterling. This offer he refused, but agreed to join them in rebuilding the engine and working the mine. The disputes which at that time arose between England and America, and the consequent revolutionary war, put an end to the projected works; and the deranged state of the country previous to the adoption of the Federal Constitution, in 1788, and other subsequent circumstances, occasioned the total neglect of this in every respect important mine, till the year 1793, when a company was formed, who undertook the work with new vigour. They collected, at a very considerable expense, miners and smelters from England and Germany, most of whom are now anxiously expecting employment in the mine. They sunk a shaft calculated to drain all the present shafts, and all that may in future be sunk, so that all the ore which may be found one hundred and fifty feet below the present levels, may be worked out without sinking any new shafts. They completed a level one hundred feet below the surface at the engine-house, which will drain the works into Hackinsack marsh, and render the lift of the engine one hundred feet less, thereby increasing its power in proportion. They repaired and improved the steam-engine so as to make it fully adequate to draining the mine and raising the ore. They purchased the fee of a tract of land called *Soho*, near and convenient to the mine; upon which they erected dams, furnaces, stampers, shops, coal-houses, and all the machinery useful for the smelting and manufacturing of copper. They provided all the tools and utensils necessary for the business; and, as well at the mine as at *Soho*, have erected dwelling-houses for the accommodation of the miners and artificers.

Soon after the mine, at the immense expense which these preparations cost (not less than one hundred thousand dollars), was rendered fit for productive operations, the affairs of the company became embarrassed: in consequence of which there was a negociation between the owners of the mine, their lessees, and the present proprietor; the result of which was, that the original term was surrendered, and the present proprietor became the purchaser of all the improvements, utensils and machinery at the mine, and of the premises and works at *Soho*, and obtained leases of the mine for thirty-three years and nine months from the sixth day of June last, at a rent of one-tenth of the ore produced by the mine, deliverable to the lessors at the works at *Soho*.

Though only a small part of the mine, in proportion to what is conceived to be its extent, has been explored, yet an inspection of the work, so far as it has been prosecuted, affords grounds for a confidence, that the veins which have hitherto proved so rich and fertile, will not fail either in the quantity or quality of their ore. But of the prospect of success and profit in the prosecution of this mine, little ought to be said on speculation. Facts only, which may easily be examined and verified, should be brought forward. It may, perhaps, be proper to state a few of these.

1. The ore of the Schuyler's mine yields, in every hundred pounds of copper, from four to seven ounces of silver, and, like most copper ores, a small portion of gold. At the time when *pure copper* was sold in England at seventy-five pounds sterling per ton, the *ore* of Schuyler's mine was shipped for England, at New-York, at seventy pounds sterling per ton. This proves the uncommon richness of the ore, and the small expense of converting it into metal.

2. The company established in 1793 have raised no ore from new ground. Their works have only been preparatory. They cleared old levels, sunk new shafts, and provided for the complete drainage of the mine for twenty years to come, though one hundred hands should be employed in it during that term. In removing the deads or rubbish which choaked the drifts and levels, they have, however, selected, without scrupulous examination, a quantity of inferior ore, which yields an average of fifteen pounds of copper, at least, in one hundred pounds of ore. This ore, which forms part of the stock on hand, will, even at so low a calculation, yield fifty tons of copper. It is asserted, from good authority, that all the deads will yield more copper per cent. than many mines profitably wrought in Germany.

3. This engine, when set to work in 1793, before the very important improvements which it has since received in its construction, and, consequently, in its powers, drained all the works dry in four days from the water which had accumulated since the last working, many years before. This proves the sufficiency of the engine, and that the objection that the mine cannot easily be drained, is quite unfounded. As to the expense of keeping the engine in repair and in operation, it will not be more than four thousand five hundred dollars per annum.

4. It has often been said, and confidently, that the first adventurers in Schuyler's mine made their fortunes, because the ore was then richer and more easily procured than at present. The fact is, however, otherwise. The ore never was more rich than lately, yielding on an average seventy-five per cent. and it will be more easily and cheaply procured in future, in consequence of the preparatory works above mentioned, and the method of working to be adopted, which will be better than has been used ever since the year 1750, when the ore retired to a considerable depth and distance from the side of the hill. And yet, since that time, while it was in its worst state of working, without an engine, and without a draining shaft lower than the levels,

four hundred and forty thousand dollars, or one hundred thousand pounds sterling, have been offered for the mine by a company of foreigners intelligent in the business.

It is true that the last company have expended large fortunes in their works at the mines, without benefit. But without going further, they did not expect profit: they had not even smelted any considerable quantity of the ore selected from the rubbish. The death of one of this small company, the return of the most active of them to England, and the very low price of copper in the European, and, consequently, in the American market, deranged and delayed their operations. At this time the Anglesea mine depreciated, by its most extraordinary fertility, every other in Europe; its proprietors were emboldened to attempt the annihilation of rivalship, by underselling every competitor; to purchase forsaken works; and thus to monopolize the copper trade of the civilized world. But the treasure disappeared almost as suddenly as it had been discovered; and the government of Great-Britain has not only been driven to prohibit the exportation of copper, but application has been made to contract for all the ore which the Schuyler's mine can produce.

A number of gentlemen, who have confidence in these facts, and who have considered that the mine still retains the benefit of the very large sums of money which have been expended merely in preparations for working it, and that, on this account, its operations must be recommenced under important advantages, have obtained from the legislature of New-Jersey, the act of incorporation which is prefixed.

It is calculated that an advance of ten dollars at the time of subscribing the incorporative deed will be sufficient to purchase the lease of the mine and the premises at Soho, and for other immediate purposes; and that a second payment of five dollars is as much as will be required for the space of six or twelve months; and it is expected that not more than five dollars further on each share will be called for till the end of the year 1803; and as the mine, long before that time, must be in full operation, it is probable that no additional advances will be wanted. Already a contract has been made with the United States for the delivery of a large quantity of copper at a liberal price, of which the benefit may, upon a small consideration, be made to result to the future stockholders.

For the purpose of carrying the act into effect, the following instrument is offered for the signature of those who are willing to become members of the corporation.

The Soho Company.

Whereas the legislature of the State of New-Jersey, in and by an act, entitled, "An Act to incorporate the persons therein named, and their associates, under the name of *The Soho Company,*" among other things, did enact, that certain persons in the said act named, together with all others who shall become subscribers for the purposes of the said act, should be, and they are thereby, together with their assigns and successors, established and made a body corporate and politic, for the purpose of conducting the operations of a copper-mine and metal company, and for no other purpose whatever, by the name and style of *The Soho Company,* and are thereby declared to be a body corporate during the term of fifty years; provided that the whole estate of the said corporation shall not exceed, at any time, the sum of half a million of dollars, exclusive of the ores and metals which they may possess and occupy under any actual productive operation. And whereas, by the same act, it is further enacted, that the capital stock of the said company shall be divided and holden in shares, so that fifty dollars of such stock shall constitute one share, as in and by the said act will more fully appear.

And whereas Nicholas I. Roosevelt hath leases for thirty-three years and nine months form the sixth day of June last, of the copper-mine, in the State of New-Jersey, commonly known by the name of *Schuyler's Copper-Mine;* and hath also the fee-simple of a certain tract of land called *Soho,* situated on Second River, near Newark, in the State of New-Jersey; which said leased premises, and tract called *Soho,* together with all the works, shafts, engines, buildings, and utensils, furnaces, stamping and boring works, coal-houses, hammers, dwelling-houses, and all and singular the appurtenances to them respectively belonging, are estimated and admitted to be of the value of sixty-five thousand dollars.

Now, therefore, we whose names are under-written, do, by these presents, declare ourselves, and do become subscribers for the purposes of the said act; and do hereby take and subscribe such number of shares in the joint stock of the said company or corporation as are set against our respective names. And we do each for himself promise and agree with all the other subscribers, that we will, at the time of subscribing, pay to the Cashier of the Bank of New-York, for the use of the said company, the sum of ten dollars on each and every share so by us subscribed. Provided always, and it is hereby agreed to be the true intent and meaning of these presents, and of all the parties hereto, that out of the said subscription monies, the said Nicholas I. Roosevelt shall be first paid the said sum of sixty-five thousand dollars.

And the said Nicholas I. Roosevelt doth hereby, for himself, his heirs, executors and administrators, covenant, promise and agree to and with all the other parties hereto, their heirs and assigns, and to and with the *Soho Company*, that in consideration of the said sum of sixty-five thousand dollars, and upon the receipt or tender thereof, he will assign to the *Soho Company* the said leases of Schuyler's mine, and all the right, title and interest of the said Nicholas I. Roosevelt, of, in, and to the same; and of, in, and to the shafts, levels, works, engines, buildings, utensils, and every other thing whatsoever belonging to or used in the said mine: and also all his estate, right and title in the said tract called *Soho,* and in all the works, dams, water-courses, races, mills, furnaces, stampers, hammers, shops, sheds, coal-houses, dwelling-houses, utensils and machinery useful and to be used for the smelting or manufacturing of copper, with all the ore now raised and lying either at the mine or at *Soho,* and all other appurtenances whatsoever. And the said Nicholas I. Roosevelt doth hereby covenant as aforesaid, that he will warrant, for himself and his heirs, the said mine and leased premises, to be by him assigned as aforesaid, for the residue of the said term of thirty-three years and nine months, and the fee-simple of the said tract called *Soho,* and all the said utensils and machinery, to the said *Soho Company.*

And whereas subscription books will be opened in the city of New-York, it is agreed that if the number of shares subscribed the first day should exceed the whole sum of ten thousand, that then there shall be a proportionate reduction from the number of each subscriber's shares above ten, so as to reduce the whole number to ten thousand.

And each of the parties, for himself, his executors and administrators, doth covenant and agree to and with all the other parties hereto, and to and with the *Soho Company*, that in case he should not pay the above mentioned sum of ten dollars on each share at the time of subscribing; and in case he should make default in the payment of any sum that may, at any time, be lawfully required of him, as a stockholder in the said company, that then the shares so by him subscribed, and all the monies that may have been paid on account of the said shares, shall be forfeited to the said company.

THE TARIFF PROTEST

Petition of the importers of Copper, in the City of New York. 25th February 1806. Referred to the Committee of Commerce and Manufactures.

To the Honorable the Senate and House of Representatives
of the United States of America.

The Memorial of the Subscribers Importers of Copper in the
City of New York Respectfully Sheweth

That by an Act of Congress passed in the Month of May One thousand Seven hundred and Ninety two, imposing duties on certain Goods Wares and Merchandizes imported into the United States it is expressly declared, That besides the exemption from duty on copper in Plates as provided for in a preeceedent Act passed in the Month of August, One thousand Seven hundred and Ninety, Copper in Pigs, and Bars should also be free from duty. That certain of your Memorialists having been for some years past in the practice of importing from Europe Copper in Bars, usually denominated *round Copper* and principally intended for the Manufacturing of Bolts for the Ship Building have been charged by the Collector of the Port of New York with the specified duty of Fifteen per cent as is imposed on Manufactured Articles of Copper from an Idea that Copper in round bars or rods is not of the description exempted by Law. This Idea your Memorialists are inclined to think has arisen from some miscontruction of the law, or from a misconception of its object and Spirit. The design of the Legislature on this particular Subject was evidently to encourage the domestic Manufacture of all Utensils, and articles constructed of Copper, and for that purpose to furnish the Workman with the unformed or raw Material at the least possible expence, and therefore the exemption of Copper in Plates, Bars and Pigs from all duty arose from the consideration that they were the mere Material roughly prepared for the hand of the Artificer and in noways interfering with the encouragement due to the domestic Manufactures. Your Memorialists are therefore at a loss to conjecture what circumstance or quality could lead to a distinction in the Judgement of the Collector, between copper forged into flat bars, and copper forged into round bars or rods. This distinction Your Memorialists are well assured is not recognized in the Port of Philadelphia, nor to their knowledge in any other Port or District in the United States, and therefore on account of its partiality as well as for many other obvious reasons it operates much to the disadvantage of your Memorialists. On this subject they beg permission further to observe, That the round bars of copper before they can be applied to the purposes of the Ship Wright, must undergo the operation not only of cutting them into suitable lengths, but the more laborious process of heading, pointing, and frequently of reducing them in size by the forge, to answer the bore of the augur, and as considerable portions of the bars often remain after cutting off the requisite lengths, the surplus is used in Manufacturing Spikes, and other fastenings for the purposes of the Ship Builder; hence it evidently appears that the round copper bars in question are nothing more than the rough Material like ordinary Bars, and that the bolts as used by the Ship Wrights are the produce of domestic Manufacture, consequently that to charge them with a duty is to discourage and check what the Law intended to favor and promote.

Your Memorialists beg leave further to represent to your Honorable Body, That they have also been for sometime past in the practice of importing Sheets or Plates of Copper for the Construction of Stills cut into certain forms according to the delineations annexed. A proportion of these pieces are calculated for the bottoms of Stills and are denominated *Flat Bottoms* being flat plates of a circular form as represented at the Letter O, and these together with the other forms marked with the Letters A B C D E F G have not been considered by the Collector of New York as subject to the payment of duty but another description denominated *Raised Bottoms,* and differing from the flat merely in having their edges turned up, have been regarded by the said Collector as not within the exemption of the Law, And your Memorialists have therefore been compelled to pay a duty for the Same in like manner as for Manufactured Articles. This exaction of duty your Memorialists are inclined to consider as unjustified by the intention of the Law, and in consequence of it, they have experienced much inconvenience and Loss; for as Flat Bottoms can easily and readily be converted by the Workman into raised ones, the circumstances of the latter being charged with a duty, rendered them unsaleable, without contributing in any sensible degree to the encouragement of domestic Manufactures. It may here be suggested by your Memorialists that in England from whence they are imported the raised Bottoms costs no more than the flat bottoms, and are not therefore regarded as more artificial in any material degree than slight Copper in plates or sheets. Your Memorialists therefore respectfully solicit your Honorable body to take into consideration their present Remonstrance and if upon due investigation it shall appear that the exemption which they claim is neither incompatible with the design nor sound construction of the Law, nor is any measure unfavorable to the Manufacturing Interest of our Country, that they would be pleased to cause directions to be given to the Collector of the Customs in the Port of New York no longer to demand the payment of duty for Copper in round bars, nor for Plates or forms denominated Raised Bottoms, And further Your Memorialists entreat that he may also be directed to refund to them all such sums of Money which he may have hitherto mistakenly exacted for duties on the said Articles.

> Harmon Hendricks
> John B. Dash Jn^r
> Thos. Harvey & Son
> William Adams
> Wm. Thomas & Son
> Fontaine Maury

The Subscribers Master Ship Wrights in the City of New York, do certify That the Statement in the within Memorial contained, relative to the formation of Copper Bolts is Substantially true and correct.

Eckford & Beebee
Adam & Noah Brown
Cheeseman & Brownne
C. Bergh

Pe[ti]tions of Brass Founders & others, praying for a Diminution of Duties on Copper Bolts & old Copper Imported into the United States.

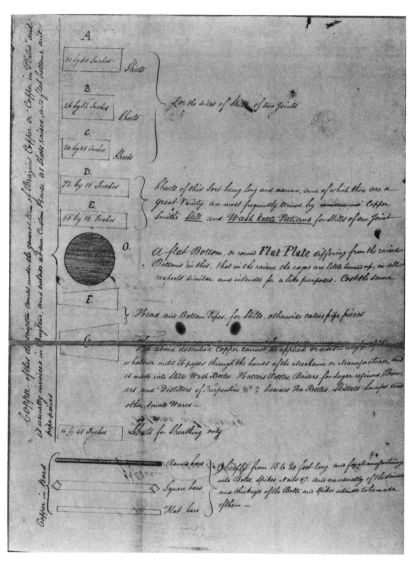

National Archives and Records Service, Record Group No. 233.

PROBLEMS OF COPPER IMPORTATION

Bristol 7 June, 1809

Harmon Hendricks Esqre

Sir

We had the pleasure of writing you 3rd May pr Packet of which we sent Copy pr Pacific: We have since been favor'd with yours of 16th 17th & 25th Ap^l & very sincerely lamenting your disappointment in not receiving your expected supply of Copper pr Alexander & Sally, who we are much Concern'd to find by your Letter had been condemned at Porto Rico on the plea of unseaworthiness arising no doubt from damage she must have receiv'd during the Voyage, whenever the papers reach us we shall use all diligence in getting the loss settled. We lament this loss the more on Acco^t of the profit which we see by your letter would, as we expected, have resulted from the due arrival & sale of the Copper. Your Bill on Messrs Savill & Sons for £48.16.7 has been duly honor'd. We were contemplating with great pleasure a renewal of our Commercial Intercourse with your Country, when to our great surprise & Concern it was officially announced to us that the Arrangements that had been enter'd into by Mr. Erskine having been unauthoriz'd by this Government, they would not for the present be confirm'd. We trust that a mutual good understanding is only suspended & that when the necessary explanations have taken place it will be fully & permanently establish'd. Until we are assured of this happy result, we do not Consider, that we can with a due regard to your Interest forward the Copp^r order'd in your favor of 25th Ap^l but which shall have our earliest attention as soon as we see that Circumstances shall justify our Shipping it.

We trust that the Copper pr Harriet & Matilda will have reached you in safety & come to a better Market than you apprehended. We shall when occasion offers Correspond with your friends Messrs Pieschell Brogden & Co. on the subject of Covering beyond the Invoice 10 pr Cent by way of profit on our future Shipm^{ts} to you as

you desire, which we Conceive may be legally effected. We observe
in your last two letters a repetition of a desire before express'd to us
that we would Confine our Shipments of Copper to your House &
your proposition as an inducement to us to give up one of the allowances of 2½ pr cent which we make you & that you would endeavor
to extend the Sum to be remitted with your Orders. We beg to recall
to your recollection the reasons which (on your former application
on this subject) we gave you in our Letter of 5 Ap¹ 1805 why we
felt ourselves Compell'd to decline acceeding to this proposal. Continuing to be influenc'd by the same motives we are under the painful
necessity of still witholding our assent to your Wishes in this respect;
& our former resolutions having been taken on a sense of the propriety
of it, we trust you will excuse us for not suffering ourselves to be
tempted to depart from it. At the same time we assure you, that we
have not been willing to give up old & respectable Connections, we
have in Compliance with your Wishes declined embracing several
flattering proposals, that have at different times been made us for
supplying new Customers in your line. Permit us to add that we do
not see how our abandoning the Correspondence of the person in
question, would as you seem to apprehend necessarily relieve you
from a Competitor in your Market; We flatter ourselves that with this
explanation, the friendly & satisfactory intercourse which has so long
subsisted between us will continue without diminution & you may
be assured that our Constant & best exertions shall be used to render
it as far as in our power agreeable & beneficial to you. We are very
sorry that in your remittances to us you should happen to have suffered at any time by any unfavorable Circumstances in the Exchange, & that in any Case it should have Counteracted the benefit
afforded by the sacrifices we have made to put you on a footing of
preference, this however, we conceived must be accidental. We hope
that the disadvantage you have sustained will be Compensated by
the Exchange proving more favorable to you in future. We are with
great respect

Sir—
Your much oblig'd & most Obed^t Serv^{ts}
John Freeman & Copper Co.

COPPER DUTIES: 1811

To the Honorable The Senate and House of Representatives of the United States.

The Petition of Harmon Hendricks of the City of New York Merchant and Samuel Denton one of the copartners of the Commercial House transacting business in the said City of New York under the Firm of Denton Little & Co. Respectfully sheweth.

That by an Act of Congress, Entitled "An Act making further provision for the payment of the Debts of the United States" passed the tenth day of August One thousand seven hundred and ninety; and by an Act, Entitled "An Act for raising a further sum of money for the Protection of the Frontiers, and for other purposes therein mentioned" passed the second day of May One thousand seven hundred and ninety two, Copper in Plates Pigs and Bars were declared to be Exempt from the payment of Duties—that your Petitioner, Harmon Hendricks for several years past has been in the practice of largely importing Round Bars of Copper commonly called Bolt Rods for the purposes of Domestic Manufacture—That the Collector of the Customs for the District of New York has heretofore required of your Petitioner on the importation of Copper Bars, Bonds for the Amount of what he alledged to be the Duties, before Permits would be granted by him for the landing of that Article—That your Petitioner, although uniformly resisting the imposition of those Duties on the ground of their not being warranted by Law, has been for several years reluctantly compelled to pay the Amount of the Bonds Exacted from him for the duties alledged to be due on the importation of that Article—And your Petitioners further shew, that by reason of their repeated remonstrances against the injustice of that imposition, and of their determination no longer to submit to an Exaction of duties wholly unwarranted by Law, the Collector of the Customs for the District of New York, for nearly two years prior to the Month of July One thousand eight hundred and nine, permitted the Bonds given by your Petitioners for the Duties on Round copper Bars imported by them

SELECTED DOCUMENTS 315

for Domestic Manufacture, to lay over, and agreed not to Exact from your Petitioners the payment of the same, or of such portion of other Bonds given by your Petitioners as related to the Duties on Round Bars of Copper, until the decision of the District Court of the United States for the District of Maryland, in which a suit on a similar Bond was then pending, should Establish their validity; Your Honorable Body having in the Years One thousand eight hundred and five and One thousand eight hundred and six determined on the application of your Petitioner Harmon Hendricks not to grant him relief, until the termination of the suit then pending in the District of Maryland—That your Petitioners sensible of the propriety and confiding in the sincerity of this arrangement with the Collector, did not hesitate to Execute from time to time, the Bonds required of them Until the month of July in the Year One thousand eight hundred and nine, when the collector of the Customs for the District of New York peremtorily and unexpectedly demanded of them the Amount due on the several Bonds which had been thus permitted by him to lay over in order to abide a Judicial decision on the point in controversy—That your Petitioners in resisting this demand urged with confidence, that the question had not received the decision of the Tribunal to which it had been referred nor of any other Tribunal in the Union; That as doubts had Existed, whether the imposition of those duties were warranted by Law, and as the suspension of payment had by Agreement between them rested solely on the ground of the illegality of those Bonds, they presumed, that no compulsory measures would be taken to Enforce the payment before a legal adjudication or a legislative interpretation had sanctioned the Legality of the imposition—But it was with much concern that your Petitioners were made Acquainted in the Month of July One thousand eight hundred and nine, with the determination of the Collector to compel the payment of these Bonds— And as Your Petitioner Harmon Hendricks had at the time goods to a very large Amount lying in the Public Stores under siezure for an alledged breach of the Nonimportation Laws, although he had every reason to believe that the siezure of those goods had been unadvisedly made, and that he was well Entitled to a restitution of them, his concern and apprehension could not be otherwise than increased when the Fate of his goods under siezure might depend on the course he should pursue in relation to those Bonds—That your Petitioners, by the advice of their Counsel paid to the Collector the amount due on their Bonds under the Express stipulation, that if it should be decided, that the Bonds were illegal, and that Round Bars of Copper, commonly called Bolt Rods were free from duty, that the amount so paid should be refunded to them by the Collector—And your Petitioners further shew that by the decision of the said Court made in the Year One thousand eight hundred and ten, the said Bonds were adjudged to be illegal and void; and that in pursuance of that adjudication, they applied to the said Collector for the whole amount paid by them on the Bonds given for Duties on Round Bars of Copper, but were informed by him that he had paid over the Money to the Secretary of the Treasury, and that an application for reimbursement must be made to them—That your Petitioners accordingly applied to the Secretary of the Treasury, who declined any interference on the ground, that the Money had been deposited in the Treasury; and recommended an Application to your Honorable Body as the only power competent to afford relief, as will appear by the Original Letter from the Secretary of the Treasury hereunto annexed—

And your Petitioners further shew, that they have been informed and believe that the practice of the Collectors in several of the Ports of the United States, has been, to debit in Account the importers of Round Bars of Copper commonly called Bolt Rods, with the amount of the duties alledged to be payable, or to adopt some other mode by which the duties might be ultimately secured in the Event of its being adjudged, that the Article was not Exempt from the payment of duties; and in reference to the Collector of the Port of New York your Petitioners would observe, that the practice of Enforcing the payment of the Bonds for the duties on Round Bars of Copper commonly called Bolt Rods, has not been uniformly and rigidly adhered to; but that, in a Case within the knowledge of your Petitioners, a demand of payment was not only protracted, but finally abandoned.

Your Petitioners therefore Respectfully pray, that your Honorable Body will Extend to your Petitioners that Relief to which they conceive themselves most justly Entitled by directing, in such Manner as your wisdom may suggest a restitution of the several sums paid by them to the Collector of the Customs for the Port of New York as herein before set forth; and for that purpose, they would beg leave to submit to the Consideration of your Honorable Body a statement, annexed to this their Petition, of the sums paid by them respectively, for the duties on Round Bars of Copper commonly called Bolt Rods imported by them since the Year One thousand Eight hundred and three;—

And your Petitioners as in duty bound will ever pray—

Harmon Hendricks
S. Denton

Statements referred to, in the annexed Petition.

Date of Importation	Names of Vessels from on Board	From what Port	Amount paid on each importation
1803 Oct 10th	Ship Penelope	Bristol	Dollars 70.00
1804 Nov 22nd	" Bristol Trader	do.	230.40
1805 Apl 30	" Oneida Chief	London	77.55
" May 13	" New York Packet	Bristol	273.80
" " 14	Brig Fair Trader	Do.	319.90
" June 27	Ship Enterprise	Do.	233.30
" Sep 1	" Venus	Do.	23.20
" Oct 3	" New York Packet	Do.	153.30
" Nov 9	" Hardware	London	197.75
1806 March 24	" Otis	Do.	152.25
May 2	" New York Packet	Bristol	374.34
May 8	" Venus	Do.	37.40
1807 Ap 25	" New York Packet	Do.	137.73
Oct 14	" Enterprize	Do.	71.75
" 28	" Shepherdess	London	469.35
Nov 7	" New York Packet	Bristol	75.60
Dec 22	" Bristol Trader	London	365.45
	Total Amount paid by Harmon Hendricks		$3263.07
	Har^n Hendricks		

1807			
Augt 18	Brig Commerce	Liverpool	271 7

Total amount paid by Samuel Denton in 3 Bonds
as follows: 1 Bond   90.35
            1 do.   90.36
            1 do.   90.36   S. Denton   271 7

This Indenture made the Tenth day of July in the Year One thousand eight hundred and Seventeen—Between James T. Joralemon of the Town Ship of Bloomfield County of Essex and State of New Jersey of the first part, and Harmon Hendricks and Solomon I. Isaacs of the City of New York Merchants of the second part Witnesseth, that for and in consideration of the [erasure] rent, covenant and agreement herein after reserved expressed and contained on the part of the lessees to be paid rendered done and performed. He, the said James T. Joralemon Hath granted, demised, set and to farm letter and by these presents Doth grant demise set and to farm let unto the said Harmon Hendricks and Solomon I. Isaacs their Executors administrators and assigns All and all manner of mines pits and veins of Copper ore, lead or other ores or other metal or metals or coal, that may be found out by digging delving sinking or otherwise howsoever lying and being the lands of the said James T. Joralemon in the said Town of Bloomfield County of Essex and State of New Jersey adjacent to and joining on the lands of Abraham Joralemons Stone Quarry and Copper Mine: with full and free liberty and licence to and for the Said Harmon Hendricks and Solomon I. Isaacs their Executors administrators and assigns from time to time and at all times during the Term hereinafter mentioned, to work and carry on the same to the best and most advantage, and to dig delve search sink trench and mine in and upon the land of the said James T. Joralemon and every and any part or parcel thereof, at their will and pleasure for the searching having and taking up of such Copper ore, lead or other ores or any other metal or metals or coal as may be there found as fully in every respect as he the said James T. Joralemon lawfully might or could do, if these presents had never been made, and the same so trenched, digged and found, to take and carry away from time to time, and at all times during the term by these presents demised or intended so to be, and sufficient ground leave for the laying of all such copper ore, lead or other ores or other metal or metals or coal as shall or may be there had or wrought, and also sufficient ground for erecting thereon such Engine or Engines, Ore House or Houses or shaft or shafts as may be necessay or requisite to be made use of in working said mines or in removing the ores that may be found on the said premises; and also all Ways, paths, passages, waters, water causes, drains, cuts, commodity, emoluments, privileges and appurtenances whatsoever thereunto belonging, or therewith to be used and enjoyed, and also, all the estate, right, title and interest of him the said James T. Joralemon to the said mines and premises; To have, hold, use, occupy and enjoy the said mines, pits and veins of Copper Ore, lead and other Ores, or other metal or metals or coal, with free liberty of digging, trenching, searching and carrying away the same, with all and singular other the premises, hereby granted, set or demised, or meant or intended so to be, with their and every of their rights members and appurtenances to the said Harmon Hendricks and Solomon I. Isaacs, their executors, administrators and assigns from henceforth, for and during and until the full end and term of Ninety-nine years from hence next ensuing and fully to be complete and ended—Yielding, rendering and paying therefor unto the said James T. Joralemon the just and full sum of Two Thousand Dollars of lawful Money of the United States one day immediately after the said Harmon Hendricks and Solomon I. Isaacs shall have commenced their operations of mining on the said demised premises with their workmen and their implements of mining; And yielding an paying therefor yearly and every year thereafter during the said Term hereby granted the annual Rent or sum of One Cent and no more—Provided always, and it is hereby declared and agreed by and between the said parties to these presents and it is the true intent and meaning hereof, that if the said Harmon Hendricks and Solomon I. Isaacs or their executors administrators or assigns shall neglect and refuse to pay to the said James T. Joralemon his heirs and assigns the said sum of Two Thousand Dollars of lawful Money of the United States of America one day after they shall have commenced mining on the said demised premises as aforesaid, that then and in that case, it shall and may be lawful to and for the said James T. Joralemon his heirs or assigns, into and upon the said mine and premises or any part thereof, in the name of the whole to reenter, and thereout and therefrom to expel and remove the said Harmon Hendricks and Solomon I. Isaacs their executors administrators and assigns, and every of them, their and every of their agents, workmen and servants, and to have and enjoy the same again as in his and their former estate, as fully as if these presents had not been made, and that from thenceforth the term estate and interest hereby granted, or so much thereof, as shall be their to come shall cease determine and be absolutely void to all intents and purposes whatsoever, these presents, or any thing herein contained to the contrary thereof in any wise notwithstanding—And the said Harmon Hendricks and Solomon I. Isaacs for themselves severally and for their several and respective heirs executors, administrators and assigns covenant grant and agree to and with the said James T. Joralemon his heirs and assigns, that the said Harmon Hendricks and Solomon I. Isaacs or one of them, or their executors administrators or assigns shall and will and truly pay or cause to be paid unto the said James T. Joralemon his heirs or assigns, the said sum of Two thousand Dollars of Lawful Money aforesaid one day immediately after they shall have commenced working or mining on the said demised premises as aforesaid, and yearly and every year thereafter the annual Rent or sum of One Cent and no more—And the said James T. Joralemon for himself his heirs and assigns, doth covenant grant and agree to an[d] with the said Harmon Hendricks and Solomon I. Isaacs their executors administrators or assigns that they or one of them paying the said sum of Two thousand Dollars at the hereinbefore mentioned and the Yearly Rent aforesaid according to the true intent and meaning of these presents, shall and may, for and during all the rest, residue and remainder of the said term of Ninety nine years peaceably and quietly have, hold, use occupy work and enjoy the said Mines pits and veins of Copper Ore lead or other ores, or other metal or metals or coal, and all and singular other the premises hereby demised or set, or meant or intended so to be, with their appurtenances, without any let, suit, trouble, interruption molestation claim or demand whatsoever of or by the said James T. Joralemon his heirs or assigns or any of them, or any other person or persons lawfully claiming or to claim, from, or under him, them or any of them—It being understood by the parties to these presents that all the Stone on the said Land which does not contain Ore of any description except such as may be necessary and requisite for the Building of the Engine or Ore House or Houses above mentioned, is reserved by the said James T. Joralemon for his own use benefit and profit—In Witness whereof the parties to these presents have hereunto interchangeably set their hands and seals on the day and in the year first above written—

Sealed and Delivered

    In the presence of   James T. Joralemon

there being an erazure from the word "the"
on the sixth line to the word "rent" on the

seventh line first page and an interlineation of "Ship" between the second and Third lines, and also an interlineation of the words "of the United States" between the Eleventh & twelth lines of the 2d page from the Top!

Samuel Davis   Harmon Hendricks

John H. King   Sol^n I. Isaacs

Docket on Face
    Deeds delivered S. I. Isaacs
    Josiah Hornblower to Mark & R 29 Augt 1794
    Jos Crane to Mark & R
    Josiah Hornblower to Mark & Roosevelt 21 Sep 1798
    James H Kip & others
    17 Nov 1813 to H H & S. I. Isaacs
    Jacob Mark & S c huyler to Nick Roosevelt 10 Sep 1799 not
      on record
    Myer Alderman to H. Hendricks release 21 Decr 1813

Crossed through
    Deeds Deliv^d S. I. Isaacs
    Oliver W Ogden—to H Hendricks & Isaacs deed for 1 octo
      1 Octr 1813
    Jno Winne to N. I. Roosevelt
    16 Augt 1799 not on record but acknowledged

CONTRACT FOR NAVAL COPPER

S. I. Isaacs & Soho Copper Compy   Copper
                with   Portsmouth, Boston, N York
Commissioners of the Navy   Philad. Wash^n & Gosport

This Contract made and entered into this thirty first day of March
Anno Domini One thousand eight hundred and twenty six between
S. I. Isaacs & Soho Copper Company of New York of the one part
and William Bainbridge, President of and acting for and in behalf
of the Board of Navy Commissioners of the United States of the other
part Witnesseth That the said S I Isaacs & Soho Copper Company
doth hereby contract and engage with the said William Bainbridge
as follow

That for the consideration herein after mentioned they will deliver
all the Copper Bolt Rods, Copper Spike rods, Sheet Copper & Composition Nails that be required from them for the use of the Navy
during the present year at the Navy Yards at Portsmouth, N.H.
Charlestown, Mass Brooklyn, N.Y. Philadelphia, Washington, & Gosport V^a to be delivered upon the requisitions of the Commandants of
the said Yards respectively—the quantities not to exceed

At Portsmouth N.H. 10 tons bolt & spike rods together—10 tons Sheet
    Copper & 3 tons Composition Nails
at Charlestown Mass. 20 tons bolt & spike rods together—10 tons Sheet
    Copper & 2 tons Composition Nails
at Brooklyn N.Y. 30 tons Bolt & Spike Rods—20 tons Sheet Copper &
    5 tons Composition Nails
at Philadelphia 15 tons Bolt & Spike rods—10 tons Sheet Copper &
    2 tons Composition Nails
at Washington 20 tons Bolt & Spike rods—10 tons Sheet Copper &
    2 tons Composition Nails
at Gosport V^a 30 tons Bolt & Spike Rods, 15 tons Sheet Copper &
    2 tons Composition Nails

all of which Copper shall be of the best and most approved quality—the Bolt Copper to be not less than ⅝ inch, the Spike rods not less than ½ inch square, the Sheet Copper to be not less than 16 oz. per square foot & the Composition Nails equal in quality and sizes to those forwarded with their offer to furnish the same to the Navy Commissioners Office—the said Copper & Nails shall be delivered at the Navy Yards aforesaid free of Expense and shall undergo the inspection thereof respectively or such other inspection as the Commissioners of the Navy may direct.

And the said S. I. Isaacs & Soho Copper Company further engage that no Officer of the Navy nor any person holding any Office or appointment under the Navy Department shall have any interest or be in any wise concerned directly or indirectly in any of the issues, profits or receipts of this Contract

And the said William Bainbridge acting aforesaid for and in behalf of the Board of Navy Commissioners of the United States doth contract and engage with the said S. I. Isaacs & Soho Copper Company as follow: that for all the Bolt Spike & Sheet Copper & Composition Nails delivered inspected approved and received as above stipulated there shall be paid to the said S. I. Isaacs & Soho Copper Company to their order by the Navy Agent residing nearest the place where the said Copper and Nails shall be delivered or at their option by the Navy Agent at New York on their presenting to the said Agent their Accounts of delivery accompanied by the certificates of the Inspecting Officers, the receipts of the Navy Storekeeper and the approval of the Commandant of the Yard at which the delivery shall be made as follow

For the Bolt & Spike Rods at Portsmouth N.H. 29 cts per pound
For the Sheet Copper & Composition Nails at Portsmouth N.H. 30 cts per pound
For the Bolt & Spike rods at Charlestown Mass 28¾ cts per pound
For the Sheet Copper & Comp^n Nails at Charlestown Mass 30 cts per pound
For the Bolt & Spike Rods at Brooklyn NY 28 cts per pound
For the Sheet Copper at Brooklyn NY 28¼ cts per pound
For the Composition Nails at Brooklyn NY 29 cents per pound
For the Bolt & Spike rods at Philadelphia 29 cents per pound
For the Sheet Copper at Philadelphia 29¼ cents per pound
For the Composition Nails at Philadelphia 29 cents per pound
For the Sheet Copper at Philadelphia 29¼ cents per pound

For the Composition Nails at Philadelphia 29 cents per pound
For the Bolt & Spike Rods and Composition Nails at Washington 29
cents per pound
For the Sheet Copper at Washington 29½ cents per pound
For the Bolt & Spike rods at Gosport Vᵃ 28⅞ cents per pound
For the Sheet Copper at Gosport Vᵃ 29½ cents per pound
For the Composition Nails at Gosport Vᵃ 29 cents per pound

And the said S. I. Isaacs & Soho Copper Company doth further engage and contract that no member of Congress shall have any interest or be in any wise concerned either directly or indirectly in any of the issues profits or receipts of this contract.

In testimony of all which agreements & stipulations the parties above named have hereunto signed their names and affixed their Seals this thirty first day of March Anno Domine One thousand eight hundred & twenty six.

Signed Sealed & delivered   S. I. Isaacs & Soho Copper Co   SEAL
in presence of   J.K. Paulding   Navy Agᵗ   SEAL
G. L. Storer   for W. Bainbridge

Naval Documents, National Archives Record Group No. 45, 340–342.

BUILDING A STATIONARY ENGINE

Estimate Engine

Mohawk & Hudson Rail Road.　　　To Uriah Hendricks, New York.
Specification of Materials &c to be used in the Construction of a Stationary Engine to be erected at the head of the Albany Inclined plane, on the Mohawk and Hudson Rail Road for the purpose of operating on said Plane, to be erected and completed by the 10th October next.

DOUBLE ENGINE

Having two Cylinders of eleven inches diameter—Cranks at right Angles Pistons 15 in. or 2 ft 6 in. Stroke. Metalic packing with Springs, Slide valves Composition. All the Journals or bearings to run in brass boxes. Cylinder piston rod glens, Composition. Valve rods and force pumps same stroke as Cylinders, plungers 2 in. diameter to supply boiler with water. Two extra pumps to raise Water to the second story. There will be required a journal on each end of the Cross head shafts for the four force pumps. Steam pipes to be made of Copper—Force pumps do—Throttle valve composition. A line of Shafts from the Engine to the upright Shaft in the centre of the Rail-way at the head of the plane.

The Company own and posess, all the *patterns* belonging to the Engine, except the bed plates and Cylinders. It may be advisable, and even *necessary*, that the builder should come, or send some competent Mechanic, and examine the Engine and apparatus, now in use, at the head of the Schenectady Plane.

Agents Office June 5th 1840

INDEX

ABOUT THE AUTHOR

Maxwell Whiteman began his career in the antiquarian book trade. His years of manuscript investigation and the study of old books led him to archival and library work. Two diverse areas attracted his interest—ethnic and minority history and early American trade, manufacture, and economics. These subjects were neglected twenty-five years ago, and Mr. Whiteman was a pioneer in bringing them to the forefront in American publishing.

Mr. Whiteman began free-lance writing and professional research in ethnic and minority history in 1949 and in 1955 joined the Hebrew Union College-Jewish Institute of Religion as assistant to the director of the American Jewish Archives and lecturer in American Jewish history. He was librarian and lecturer in American Jewish history at the Dropsie College for Hebrew and Cognate Learning from 1958 to 1964.

Since 1964 Mr. Whiteman has been archival and historical consultant to the Union League of Philadelphia, where he oversees its magnificent library of Civil War literature and general American history.

During the years of Mr. Whiteman's institutional associations, he has continued to pursue his first interest, the search for old and curious books. He has traveled throughout the United States and Canada, the continent of Europe and the Middle East, and believes that there is special action and excitement in the quest of the antiquarian, both in the objects sought and in the people encountered.

Mr. Whiteman is the author of *A Century of Fiction by American Negroes* (1955), *A History of the Jews of Philadelphia from Colonial Times to the Age of Jackson* (with Edwin Wolf 2nd), 1957, *Mankind and Medicine: A History of Philadelphia's Albert Einstein Medical Center* (1966), *Pieces of Paper: Gateway to the Past* (1967), *While Lincoln Lay Dying* (1968), and he edited the forty-six-volume Afro-American History Series (1969), and *The Kidnapped and the Ransomed: The Narrative of Peter and Vina Still After Forty Years of Slavery*, with an introductory essay on Jews in the antislavery movement (1970).

The text of this book was set in Caledonia Linotype and printed by offset on P & S Special XL manufactured by P. H. Glatfelter Co., Spring Grove, Pa. Composed, printed & bound by Quinn & Boden Company, Inc., Rahway, N.J.